A COMPREHENSIVE EXPLORATION OF THE SCIENTIFIC MIRACLES IN HOLY QURAN

MAHDI LA'LI

Order this book online at www.trafford.com
or email orders@trafford.com

Most Trafford titles are also available at major online book retailers.

© Copyright 2007 Mahdi La'li.
All rights reserved. No part of this publication may be reproduced,
stored in a retrieval system, or transmitted, in any form or by
any means, electronic, mechanical, photocopying, recording, or
otherwise, without the written prior permission of the author.

Print information available on the last page.

ISBN: 978-1-4120-1443-4 (sc)
ISBN: 978-1-4269-3220-5 (e)

Because of the dynamic nature of the Internet, any web addresses or links contained in
this book may have changed since publication and may no longer be valid. The views
expressed in this work are solely those of the author and do not necessarily reflect the
views of the publisher, and the publisher hereby disclaims any responsibility for them.

Any people depicted in stock imagery provided by Getty Images are models,
and such images are being used for illustrative purposes only.
Certain stock imagery © Getty Images.

Trafford rev. 08/08/2018

 www.trafford.com

North America & international
toll-free: 1 888 232 4444 (USA & Canada)
fax: 812 355 4082

*In the Name of God,
the Most Gracious, the Most Merciful*

And We have not sent thee, save as a mercy unto all beings. Chapter 21: verse 107

Dedication

With much love and honor, this book is written in dedication to the holy prophet of mercy, the light of guidance and truth, the last messenger of God, Muhammed, may peace and blessings be upon him.

Acknowledgements

My appreciation is extended first of all to Hassan Fatemian, who caused me to stretch and view life through another window. My thanks go to him not only for his contribution in graphic designing, but also for encouraging me in memorizing the Quran over the course of years.

I am profoundly grateful to my esteemed sister in Islam, Ann Marie (Zeinab) Yezzi Shareef, a professor of Gannon University, Pennsylvania USA, who reviewed and edited all the articles of this book with her exquisite virtuosity. Her vast educational background and useful recommendations were indeed effective in accomplishing this work. My zeal has been sparked by her patience and sincerity to pursue this aim and complete it. I needed her.

I especially owe a great debt of gratitude to other brothers and sisters for the assistance I received in researching and providing the scientific sources, editorial reviews, rewriting, typing, scanning, proofreading, graphic designing, title and the honest feedback. For this, I thank Dr. Anita Eftekharzadeh, Rose Starratt from Nova Scotia Canada, Seyyed Muhammed Atabaki, Ali Fatemian, Ehsan Mazidi, Muhammed Javad Esmaeli, Hadi Alimardani, Hossein Fatemian, Muhammed Javad Mouti'yan and Razieh Parastan.

Finally, I would like to say a special thank you to all the members of Trafford publishing staff for offering wonderful services and giving back to the authors.

I am exceedingly obliged to thank all of those who have helped me and to wish them the best of bounties from God Almighty and pray with all my heart that He may reward them here and in the hereafter. Amen

Contents

Preface 17
Introduction 19

Cosmology in Holy Quran 23

Big Bang Theory 25
Scientific Viewpoint 26
    Distant Galaxies 27
    Particle Accelerators and Detectors 27
Quranic Viewpoint 28

Expanding Universe 31
Scientific Viewpoint 32
Quranic Viewpoint 33

Gaseous Origin of the Universe 35
Scientific Viewpoint 35
Quranic Viewpoint 36

Extraterrestrial Life 39
Scientific Viewpoint 40
Pulsar Planets 41
Quranic Viewpoint 41

Termination of the Universe 44
Scientific Viewpoint 46
Quranic Viewpoint 47

Destiny of Earth **50**
   Scientific Viewpoint 51
   Quranic Viewpoint 51

Death of the Universe **53**
   Scientific Viewpoint 54
      Death of a Star 54
      Death of an Ordinary Star 55
      Death of a Massive Star 55
      White Dwarfs 56
      Red Giant 56
      Supergiant Stars 57
      Supernovae 57
   Quranic Viewpoint 58
      A Dazzling Evidence 60
      NASA's Explanation 60

Astronomy in Holy Quran 63

Interstellar Galactic Material **65**
   Scientific Viewpoint 66
      Empty Space is Not Empty 66
      Interstellar Medium 66
   Quranic Viewpoint 68

Revolution of the Heavens **69**
   Scientific Viewpoint 70
   Quranic Viewpoint 70

Black Holes **72**
   Scientific Viewpoint 73
      Black Hole's Formation 73
      Black Holes, Gravity's Ultimate Victory 74
      Curved Space-Time 75
   Quranic Viewpoint 75
      Dark Stars 75
      Lost Stars 76
      Piercing Star 76
      Time Dilation in Black Holes 77

Pulsars and Neutron Stars **80**
   Scientific Viewpoint 81

  What is a Neutron Star? 81
  How Neutron Stars Form 81
  Pulsars 82
  Kepler's Laws apply to binary stars also! 82
 Quranic Viewpoint 83

Sun and Moon and Their Orbits 85
 Scientific Viewpoint 86
  The Existence of Sun's Orbit 86
  Orbits of Sun and Moon 86
 Quranic Viewpoint 87
  The Sun and Moon Move With Their Own Motion 88

Inherent Diversity between Sun and Moon 89
 Scientific Viewpoint 90
  The Sun 90
  The Moon 91
 Quranic Viewpoint 91

Revolution of Sun 94
 Scientific Viewpoint 95
 Quranic Viewpoint 96

Composition of Meteors 98
 Scientific Viewpoint 99
  Meteorite Types 99
 Quranic Viewpoint 100

Motion of Earth 102
 Scientific Viewpoint 103
 Quranic Viewpoint 103

Trundling Earth 105
 Scientific Viewpoint 106
  Precession of the Earth Axis 106
 Quranic Viewpoint 107

Spherical Earth 109
 Quranic Viewpoint 110

Aerospace in Holy Quran 113

Aerodynamics and Flight Control 115

- Scientific Viewpoint 117
  - Aerodynamics 117
  - Control and stability 117
- Quranic Viewpoint 118

Spacecraft 121

- Scientific Viewpoint 122
  - Launch Vehicles 122
  - Thruster 122
  - How Rockets Work 123
  - A. Action and Reaction 123
  - B. Thrust and Efficiency 123
  - Staging 124
  - Rocket Flight 124
  - Stability and Control 125
  - Gravity 125
- Quranic Viewpoint 126

Conquest of Space by Human 128

- Scientific Viewpoint 129
  - Science of Space Exploration 129
- Quranic Viewpoint 130
  - Getting into Space 131
  - Escape Velocity 132

Physics in Holy Quran 135

Universal Gravitation 137

- Scientific Viewpoint 138
  - Gravity 138
  - Electromagnetic Force 139
  - The Strong Nuclear Force 139
  - The Weak Interaction 139
- Quranic Viewpoint 139

Gravity of Earth 141

- Scientific Viewpoint 142
  - Newton's Principles of Mechanics 142
- Quranic Viewpoint 143

Universality of Pairing 145

- Scientific Viewpoint 146
- Quranic Viewpoint 146

The Well-Balanced Universe 148
Scientific Viewpoint 151
- Cosmythology 151
- Was the universe designed to produce us? 151
- Tuned For Life 151
Quranic Viewpoint 153

The Origin of Iron 156
- The Ancient Finnish Myths about the Iron 157
- What is the origin of Iron? 157
Scientific Viewpoint 158
Quranic Viewpoint 159

Water and Its Individual Properties 161
Scientific Viewpoint 162
- What Is Water? 162
- Three phases to a quick-change artist 163
- Forming the link to life 163
- Powerful Properties Shape Our World 163
- World's best dissolver 164
- Solid expansion 164
- A heated exchange 164
- Tension on top 165
- Sticky sides 165
- Global recycling 165
- Ocean Commotion 166
Quranic Viewpoint 166

Meteorology in Holy Quran 169

Protective Properties of Atmosphere 171
Scientific Viewpoint 172
- Ozone Layer 172
- Formation of the Ozone Layer 172
Quranic Viewpoint 173

Drop of Atmospheric Pressure 175
Scientific Viewpoint 176
- Scientific facts concerning the state of man in high altitude 176
- Symptoms of the stages of Hypoxia 177
- Drop in Atmospheric Pressure 177
Quranic Viewpoint 178

Seven Layers of Atmosphere 179
    Earth's Atmosphere 179
    Atmospheric Layers 180
  Scientific Viewpoint 180
    Seven Layers of Atmosphere 180
    Troposphere 181
    Stratosphere 181
    Ozone layer 181
    Mesosphere 182
    Ionosphere 182
    Thermosphere 183
    Exosphere 183
  Quranic Viewpoint 184

Reduction of Atmosphere 185
  Scientific Viewpoint 185
    Historical Atmosphere 185
    Exosphere 186
  Quranic Viewpoint 186

Formation of Clouds and Rains 188
  Scientific Viewpoint 189
    Why is the sky blue? 189
    And Clouds are White Because…? 189
    Why do clouds form? 189
    Formation of Clouds 190
    Lightning 190
    Why Does It Rain? 190
  Quranic Viewpoint 191

Medical Sciences in Holy Quran 193

Embryology 195
  The Creation of Man 195
  Scientific Viewpoint 196
    First to Third Week 197
    Fourth to eighth week: Embryonic Period 198
    From Third to Tenth Month: Fetal Period 199
  Quranic Viewpoint 199
    Skeptic's Reaction 200
    A Scientist's Interpretation of Embryology in the Quran 201
    The Embryonic Phases 203

The Three Veils around Fetus 207
Scientific Viewpoint 208
Quranic Viewpoint 209

Gender Determination 214
Scientific Viewpoint 215
Formation of Male and Female Sex Cells 215
Fecundity 216
Quranic Viewpoint 216

Nature of Sperm 218
Scientific Viewpoint 219
Quranic Viewpoint 219

The Sensory Characteristic of Skin 221
Scientific Viewpoint 222
What is Skin? 222
Quranic Viewpoint 222
Miracle in the Quran 223

Cerebrum 226
Scientific Viewpoint 227
Cerebrum 227
Anatomy of the forehead 227
Scientific Signs in human behaviour with regard to the lobe 228
The Lobe and the higher mental functions 228
Quranic Viewpoint 229
Scientific signs in the Quran and Sunnah with regard to the Lobe .. 229
Scientific signs with regard to the Lobe 229

Fingerprints 231
The History of Fingerprints 232
Why Fingerprint Identification? 235
Scientific Viewpoint 236
Quranic Viewpoint 237

Therapeutic Properties of Honey 238
Scientific Viewpoint 239
Quranic Viewpoint 240

Menstruation 242
Scientific Viewpoint 243

    Menstruation...243
    Harmful Effects of Coitus on Women at the Time of Menstruation ..244
  Quranic Viewpoint 245

Geology in Holy Quran 247

Geological Description of Mountains 249
  Scientific Viewpoint 251
    Mountains, the Skeleton of Earth 251
    Mountains, the mighty claws 251
    Mountains, the regulators of the winds.......................................251
    Mountains, the Natural Refrigerators..252
  Quranic Viewpoint 252

Lowest Land of Earth... 256
  Scientific Viewpoint 257
    Dead Sea, the lowest point on earth..257
  Quranic Viewpoint 258

Oceanography in Holy Quran 261

Oceans and the Internal Waves 263
  Scientific Viewpoint 264
    Internal Waves 264
  Quranic Viewpoint 265

Zoology in Holy Quran 269

Intelligent world of Animals 271
  Scientific Viewpoint 272
    Animal Behavior 272
    Programmed Learning 274
    The Ant 275
    Honeybee 278
    A. Reproduction and Development 280
    B. Activities 280
    C. Communication 280
    D. Problems of Survival 281
  Quranic Viewpoint 283
    The Female Bee in the Quran 284

Bibliography 286

Preface

The Most Glorious and Holy Quran as the last revelation of God Almighty, is the perpetual miracle of Muhammed (peace be upon him), the prophet of Islam, who presented it to all nations of all time over 1400 years ago. Since, ours is an era of amazing scientific discoveries, to corroborate the divinity of the Quran by science would be a strong and undeniable proof for the human of the contemporary epoch. Logically, the existence of such newly discovered scientific facts in Holy Quran inspirationally indicates that it could certainly have not had a human source and is no less than a preternatural revelation. The most Holy Quran is indeed a gift to mankind across all ages and all times.

Since my adolescence, the verses, which clearly pointed out the physical facts, always sparked my interest. I used to dwell upon them and find the relation between them and science. After graduation from the university, I found the time to explore the scientific aspects of the Quran in detail.

In the course of the past years, I have attempted to clarify some of the verses, which are explicitly in relation with various scientific facts. In the present book, any article comprises three main parts: Introductory comment, Scientific Viewpoint and Quranic Viewpoint. I have tried to prepare the most creditable and newest scientific sources as the evidence in regard with the respective Quranic statements. In addition, the best English translations of Holy Quran are presented to provide the readers with a vivid comprehension of the Quranic purports.

At the end of any article, the sources of scientific and Quranic notions are annotated as the references, which enable the readers to verify the documentations. Some Articles, also, include "Further Reading" to supply the readers with additional sources for more investigation.

An extra attention has been paid on accuracy in this book. Before publishing, I have made certain that the texts have been reviewed by many of my friends for clarity and avoiding any mistake of any kind. Their contributions have been so helpful.

A significant attempt has also been made in simplifying the book and making it easy to read and study from. The scientific viewpoints are tried to be conceptually generalized without any formulas or the specialized information so that they are intelligible for any group of readers with any kind of education and level of knowledge.

My main intention in writing this book in English has been to globalize it, in such a manner that it can convey its message of scientifically proving the Holy Quran's divinity and authenticity, especially to those non-Muslims who are unaware of God Almighty's final testament, which offers the perfect guidelines to humankind.

Since 'to err is human', mistakes may have been made. Therefore, I would really appreciate it, if you could inform me of your valuable opinions in this issue. Obviously, any useful suggestion and constructive criticism is certainly welcomed and highly appreciated.

In finality, it is necessary to mention that we have not meant to claim that our viewpoints are exactly what God Almighty has intended to ordain. This book is the author's own presumption, which is based on the contemporary science. The future scientific discoveries may declare some other facts, which will reveal other aspects of the related issues. Nonetheless, it is the author's contention that Holy Quran will indeed hold up to the test of time and future discoveries.

Mahdi La'li
omidmed@hotmail.com
Teheran, November 2002

Introduction

All praise belongs to God, the most Compassionate, and the most Merciful; and may peace and blessings be upon His last Messenger, Muhammed.

No book has affected the human communities as much as the Holy Quran has; and this affection has been conspicuous from the incipient eras of Islam. The fast transition and marvelous revolution, which occurred among the Arab communities and the nation of Islam, has been unique among all societies of all time. This revolution was undoubtedly because of the amazing power contained in the revelations of Holy Quran.

God has sent numerous prophets to all the nations and races. As the human race progressed, the prophets arrived with the laws that suited the requirements of that time. Each new prophet brought a new divine law, which abrogated or cancelled the previous law. Muhammed is the last prophet of God and he has brought the last and the most perfect law in the Holy Quran. History shows us that this law has suited the requirements of the people for the last 1400 years and shall continue to do so till the day of resurrection.

The prophets were sent by God with their own miracles to prove their prophecy to the people. These times have passed and with them the prophets and observable evidence of their miracles. Namely, their miracles have not been perceivable by other nations in other places and eras. For example, man can never see Moses again converting his staff to a serpent before Pharaoh, or Jesus Christ raising the dead by God's leave.

The only tangible miracle that yet remains in our world today is the Holy Quran, God's conclusive word to this world. Islam, as the final and the everlasting religion of God, has been charged to offer this perpetual miracle in such a manner that it would have been accessible for the people at all times.

What is the miracle of Muhammed? Various miracles such as splitting of the moon, food multiplication, supplication for rain, the prophet's night journey from Medina to Jerusalem and ascent to the heavens etc, have been narrated and some are included in Holy Quran. However, during the past times, Muslims have believed in the prophecy of Muhammed through his perpetual miracle i.e. Holy Quran.

From many different aspects, Holy Quran is deemed to be a miracle. It is a wonderful piece of poetry and unique Arabic literature, full of wisdom and guidance. On reading it, one is at once convinced that it is the very Word of God, for no man can write such a perfect guidance on so many subjects with such eloquence and charisma. In all respects it supercedes what any human could have compiled.

One of other miraculous aspects of the Quran is the existence of modern scientific data in it. This information could have not been easily conceived considering the time and the place in which the Quran was revealed. Muhammed was an illiterate man who never studied, wrote or read anything and was raised in the most backward nations in his time. Nevertheless, numerous scientific notions, which have been yet to be discovered in the intervening course of history, are mentioned in the Quran. Expansion of the universe, gravity of earth, universal gravitation, oceans' internal waves, fetus embryonic phases, protective properties of atmosphere, reduction of atmosphere, lowest land of earth, composition of meteors, black holes, pulsars etc, being clarified in an ancient book, is not a matter to be easily neglected.

The book of God constantly appeals to one to think, ponder and understand; and forbids one to drown one's reasons or believe blindly. God says in Holy Quran:

-Surely in the creation of the heavens and earth and in the alternation of night and day there are signs for men possessed of minds; [those] who remember God, standing and sitting and on their sides, and reflect upon the creation of the heavens and the earth: Our Lord, Thou hast not created this for vanity. Glory be to Thee! Guard us against the chastisement of the fire.

Chapter3: verse190 and 191

Holy Quran mentions that no man will be able to forge even a part of it and that no corruption shall touch it from any side. In fact, it challenges any man to do so. It is also a miracle that Holy Quran has remained unchanged or altered an iota during all these 1400 years. It shall remain so until the final Day of Judgment, for God has taken it on Himself to protect it.

Another miraculous aspect of the Holy Quran is its statistical wonders. As an example, the word "world" is mentioned 115 times in the Quran and this interestingly equals the number of times that the word "hereafter" is used as well. Other examples are "angels", "demons" each mentioned 88 times, life and death, 145 times, and man and woman each used 24 times equally. Also the word "sea" is mentioned 32 times and the word "land" is used 13 times in the Quran. If we add up the total words of both "sea" and "land" we get 45. Now if we do a simple calculation:

$32/45 \times 100\% = 71.11111111\%$ seas
$13/45 \times 100\% = 28.88888888\%$ lands

Above is exactly what we know today about the percentages of water and land, which cover the surface of earth. This is yet another miracle in the Quran!

Also, it is wonderful that the number of word "day" mentioned in the Quran, equals the number of days of the year, that is, 365. Similarly, the word "month" is mentioned 12 times in the Quran, which is the number of total months of the year.

The Miracles of the Quran will never end. God has challenged in the Quran that no one will ever be able to compose a book like it:

-Say: Verily, though mankind and the Jinn (demon) should assemble to produce the like of this Quran, they could not produce the like thereof though they were helpers one of another.

Chapter17: verse88

Somewhere else, man is challenged to produce 10 chapters the like of the Quran's:

-Or do they say, 'he has forged it'? Say: 'Then bring you ten suras (chapters) the like of it, forged; and call upon whom you are able, apart from God, if you speak truly.'
-Then, if they do not answer you, know that it has been sent down with God's knowledge, and that there is no god but He. So have you surrendered?

Chapter11: verse13

In His final challenge, God has restricted the battlefield and called anyone to bring only one chapter the like thereof:

-And if you are in doubt concerning that We have sent down on Our servant, then bring a sura like it, and call your witnesses, apart from God, if you are truthful.
-And if you do not -and you will not- then fear the Fire, whose fuel is men and stones, prepared for unbelievers.
Chapter2: verse23 and 24

...And history vividly witnesses that no one has answered to this challenge yet.

Cosmology in Holy Quran

- Big Bang Theory
- Expanding Universe
- Gaseous Origin of the Universe
- Extraterrestrial Life
- Termination of the Universe
- Destiny of Earth
- Death of the Universe

Big Bang Theory

أَوَلَمْ يَرَ ٱلَّذِينَ كَفَرُوٓا۟ أَنَّ ٱلسَّمَٰوَٰتِ وَٱلْأَرْضَ كَانَتَا رَتْقًا فَفَتَقْنَٰهُمَا ۖ وَجَعَلْنَا مِنَ ٱلْمَآءِ كُلَّ شَىْءٍ حَىٍّ ۖ أَفَلَا يُؤْمِنُونَ ۝

The world we live in is a marvelous collection of wonders, which amazes any enlightened beholder. It makes him admire its majesty by its dazzling greatness and precise reckoning.

How did this wonderful universe with such great dimensions come to be in the existence? When did it happen? Is it going to die one day? Currently, the Big Bang model of the origin of the universe is the cosmological paradigm most widely accepted by astronomers. This theory contends that about 15 billion years ago the universe began with the explosive expansion of a single, extremely dense matter, the primordial mass. Astronomers arrived at this conclusion only after the development of radio telescopes in 1937 and the necessary observational precision was achieved.

The Holy Quran was written over 14 centuries ago, as it was revealed to the Prophet of Islam, Muhammed (PBUH). One of the glorious mysteries of the Quran is that it explains many of the mysteries of our universe, long before modern scientists could reveal these same secrets. What does Quran say about the origin of our universe? Wonderfully enough, the following verse indicates that the universe was first a unit piece then the heavens and the earth originated from it:

-Have not those who disbelieve known that the heavens and the earth were of one piece, then We parted them... will they not then believe?[1]

Chapter21: verse30

In the following passage, we will offer you the Quranic overview regarding the origin of the universe, which is quite basically in accordance with the new scientific discoveries. To achieve a better contemplation of the corresponding Quranic viewpoint, we should first glance over the pertinent scientific theories.

Scientific Viewpoint

How do modern scientists explain the formation of the universe? Dr. Maurice Bucaille explains it in his book 'The Bible, the Quran and Science':

"The basic process in the formation of the universe ... Lay in the condensing of material in the primary nebula followed by its division into fragments that originally constituted galactic masses. The latter in their turn split up into stars that provided the sub-product of the process, i.e. the planets".

Astronomers postulate that about 15000 million years ago, all the material in the universe was sent flying outwards in all directions. They can not explain how or why this happened, but they compare it with a massive explosion and have labeled this 'The Big Bang Theory'[2]. The big bang model is currently, the only widely accepted explanation for the origin of the universe.

The best evidence for the Big Bang Theory is the existence of cosmic microwave background radiation. This is very faint microwave radiation, which has a long wavelength and which fits with the earlier discovery by American astronomer Edwin Hubble (1889-1953) that the universe is expanding.

This cosmic microwave radiation also confirms the prediction by Ukraine-born American physicist George Gamov (1904-1968) who proposed the theory that if there was a beginning of the universe, then the radiation now reaching us from it, would be coming from the furthest parts of the universe, which have been rushing away from us at great speed. His theory suggested that such radiation would have been subjected to the most extreme red shift. He defined a red shift as a measure of the degree to which light from a receding object moves toward the red end of the electromagnetic spectrum and would therefore be expected to contain only long wavelengths.

Distant Galaxies

Further evidence for the Big Bang Theory is provided by the study of distant galaxies. Some of these galaxies are 13 billion light years away, which means that the light we see from them will have taken 13 billion years to reach us. These Galaxies appear to us, as they would have appeared around two billion years after the Big Bang. The fact that they seem more tightly packed than closer galaxies would indicate that the universe is increasing in volume as it gets older and was once much smaller and denser. In other words, if the light from more proximate galaxies reaches us at a faster rate, and they appear to be less condensed, therefore, the "older picture" of distant galaxies gives us the picture of how things were. Comparing the nature of the distant galaxies to those who we are seeing in the present time we can see the galaxies are expanding from their original formation.

Particle Accelerators and Detectors

Scientists attempt to re-create the conditions that existed immediately after the Big Bang in the hope of uncovering the origins of the universe. They do this by sending two beams of subatomic particles in opposite directions around an instrument called a particle accelerator (a device used for detecting subatomic particles), causing the beams to collide as they approach the speed of light.

The energy of the collision creates new particles. These particles leave trails in a bubble chamber (a device where subatomic particles travel through liquid hydrogen, causing it to boil and leave trails of bubbles) enabling scientists to identify them. The results of these experiments reveal much about the early universe, because the energy of the colliding particles is similar to the energy of particles in the first moments after the Big Bang.

At the Conseil European pour la Recherché Nucleaire (CERN) near Geneva in Switzerland, there is a cyclotron (a circular particle accelerator) in a huge underground tunnel. The circumference of the outer edge of the tunnel measures about 27 km (17 miles)[3]. It is here that further investigations are revealing new facts regarding the creation of the universe.

Whatever happened in the initial moment of the Big Bang remains a mystery, but it is likely after a tiny interval, the universe was still smaller than the nucleus of an atom, called *the primeval atom*[4]. Almost immediately, the strong nuclear force and the electromagnetic force separated, providing the energy needed for the universe to start inflating. The Big Bang itself was extremely hot and energetic and in the first few seconds after the explosion, all that existed in the universe was radiation and various subatomic particles. Before even a millisecond had past, the universe had grown to about 3 times the size of the sun. Continued inflation caused the universe to double in size every fraction of a second.

The radiation left over after this vast explosion is still detectable from earth as faint microwaves emanating from all directions of the sky. This is the cosmic microwave background radiation first reveled by Hubble and Gamov discussed earlier.

Quranic Viewpoint

Does the Quran say anything about this condensing and separation of the primary material to result in the formation of our universe? Let's have a look. Our creator, Allah, says in his final book:

-Do not the unbelievers see that the heavens and the earth were joined together, then we clove them asunder.

Chapter21: verse30

This could also be translated as follows:

-Do not the unbelievers see that the heavens and the earth were fused together, then we separated them ...

Dr. Bucaille considers this proof of "the reference to a separation process of a primary single mass whose elements were initially fused together".

Thus, the Quran gives an accurate account of the formation of the universe to call upon humankind to recognize the power of their creator. This raises an interesting question: How could a man living in the seventh century independently create these ideas which could not be confirmed until modern times? And how could he, in so doing, avoid the mythical and fanciful ideas prevalent in human history of the period prior to scientific discovery and investigation into these concepts?

Dr. Bucaille mentions some of these myths for contrast: "When, as in Japan, the image of the egg plus an expression of chaos is attached to the above with the idea of a seed inside an egg (as for all eggs), the imaginative addition makes the concept lose all semblance of seriousness. In other countries, the idea of a plant is associated with it; the plant grows and in so doing, raises up the sky and separates the heavens from the earth. Here again, the imaginative quality of the added detail lends the myth its very distinctive character".

In contrast to those and other similar myths of pre-scientific thought, the Quranic statements are "free from any of the whimsical details accompanying such beliefs; on the contrary, they are distinguished by the sober quality of the words in which they are made, and their agreement with scientific data". (Bucaille, 1987) So what would account for this miraculous attention to true scientific findings before there was evidence and the capacity to even investigate such areas of science? There can be only one rational explanation to any enlightened mind and that is the Quran is unique in that[5]. It must be that the Quran is not the product of any human or humans, but a revelation from Allah.

The Quran says:

-The revelation of the scripture whereof there is no doubt is from the Lord of the Worlds.

Chapter32: verse2

Professor Alfred Kroner is one of the world's most famous geologists. He is a Professor of Geology and the Chairman of the Department of Geology at the Institute of Geosciences, Johannes Gutenburg University in Mainz, Germany. Professor Alfred Kroner states that: "many of the statements made at the time could not be proven, but that modern scientific methods are now in a position to prove what Muhammed said 1400 years ago".

Sheik Abdul Majid Zendani, a Professor of Islamic Studies in King Abdulaziz University in Jeddah, Saudi Arabia says:

"We met him (Dr. Kroner) and presented several Quranic verses and Ahadeeth (narrations) of the Prophet Muhammed. He studied and commented on them. Then we had a discussion with him".

Professor Kroner said: Thinking about many of these questions and thinking where Muhammed came from, he was after all a Bedouin (a desert nomad). I think it is almost impossible that he could have known about things like the common origin of the universe, because scientists have only found out within the last few years with very complicated and advanced technological methods that this is the case.

Professor Kroner chose an example from the Quran, which proved to him why the Quran could not have come from Muhammed. The example, which Professor Kroner chose, is a description in the Quran of the fact that this universe had its beginnings in one single entity. Allah, may He be Exalted and Glorified said:

-Do not the unbelievers see that the heavens and the earth were joined together [ratqan], before we clove them asunder?

The meaning of "ratqan" in this verse, as Ibn Abbas, Mujaahid, and others said, is that the heavens and the earth were stuck together or blended together, and that they were later separated from each other. Professor Kroner used this as an example to prove that no human being during the time of Prophet Muhammed could have known this.

Professor Kroner: "Somebody who did not know something about nuclear physics 1400 years ago could not, I think, be in a position to find out from his own mind for instance that the earth and the heavens had the same origin or many others of the questions that we have discussed here".

Professor Kroner becomes evasive if embarrassed. But whenever he is faced with the truth, he is courageous enough to state his opinion frankly and thus he replied: "This could have been known to him only through revelation from above". Finally, after our discussions with him, he made the following comments:

"If you combine all these and you combine all those statements that are being made in the Quran in terms that relate to the earth and the formation of the earth and science in general, you can basically say that statements made there in many ways are true. They can now be confirmed by scientific methods, and in a way you can say that the Quran is a simple science text book for the simple man, and that many of the statements made in there at that time could not be proven but that modern scientific methods are now in a position to prove what Muhammed said 1400 years ago" [6]

Most assuredly it can be shown that in the time of revelation of this verse, no one was cognizant of the origin of the universe and how it began; but today, it is apprehended that the scientific purport of this verse adapts with all new discoveries and findings about the origin of the universe.

References
1. Holy Quran translated by Marmaduke Pickthall, George Allen and Unwin Ltd., London, Fifth Edition 1969.
2. Eyewitness Encyclopedia of Space and the Universe, by Dorling Kindersley Publishers Limited, London, 1990.
3. ibid.
4. Exploration of the Universe, by George Abell, Hot, Rinehart and Winston Inc., USA, 1969.
5. The Bible, The Quran and Science (Le Bible, le Coran et la Science)," The Holy Scriptures Examined in the Light of Modern Knowledge, by Dr. Maurice Bucaille, French Physician, Seghers, Paris, 1987, English version published by North American Trust Publication, 1978.
6. Holy Quran on the Big Bang Theory, http://www.islam-guide.com/.

Expanding Universe

وَٱلسَّمَآءَ بَنَيْنَٰهَا بِأَيْي۟دٍ وَإِنَّا لَمُوسِعُونَ ۞

As previously discussed, the universe has been expanding since its creation. Hubble, an astronomer made this great discovery, in the early part of this century. In 1925, Edwin Hubble (after whom the Hubble Space telescope is named) provided the observational evidence for the expansion of the universe. Stephen Hawking (author of 'A Brief History of Time') states: "The universe is not static, as had previously been thought, it is expanding".

The galaxies other than our "Milky Way" are moving away from us at tremendous speeds. This can be ascertained, by analyzing the light given off by the galaxies, by an instrument called the spectroscope. The more distant a galaxy is, the greater is its speed of recession. This motion of the galaxies is due to the expansion of the universe. As previously discussed, the universe began about fifteen to twenty billion years ago with a tremendous explosion called "The Big Bang" and has been expanding ever since. The universe is not just a sphere of galaxies rushing away from each other into unending emptiness of space. It can be compared to a balloon with many dots on it, each representing a galaxy. However, the surface of this balloon, which is two dimensional, represents all three dimensions of our space, and is expanding into a higher dimension, that is beyond our imagination[1].

What did the Quran reveal 1400 years ago concerning this issue? Let us have a look. Astonishingly enough, the expansion of the universe is clearly mentioned in the verse 47 of chapter51:

-[And] We have built the heaven with might, and We it is who make the vast extent (thereof)[2].

Chapter51: verse47

To understand how the universe is expanding, some knowledge of cosmology is essential.

Scientific Viewpoint

So far we have learned that in the late 1920's the American astronomer Edwin Hubble (1889_1953) analyzed starlight from distant galaxies. He discovered that the wavelengths of the light were longer than expected.

It was first suggested by the general theory of relativity and is backed up by physics in the examination of the galactic spectrum; the regular movement towards the red section of their spectrum may be explained by the distancing of one galaxy from another. Thus the size of the universe is probably constantly increasing and this increase will become bigger the further away the galaxies are from us. The speeds at which these celestial bodies are moving may, in the course of this perpetual expansion, go from fractions of the speed of light to speeds faster than this.

This effect called red shift indicated that these galaxies were moving away from the origin of the Big Bang at great speed. A similar effect can be observed in all directions from earth, leading astronomers to assert that the universe is getting perpetually larger, but to what end?

When astronomers look out into space far beyond our galaxy, they see many other galaxies all seem to be moving away from us and from each other and those furthest away are moving the fastest. So we seem to be in all expanding universe, but what happened to start this expansion?

As we stated before, astronomers think that about 15000 million years ago all the material in the universe was sent flying outwards in all directions. They cannot explain how or why this happened, but they compare it with a massive explosion: the big bang. At first the universe was very hot, but it cooled as it expanded and became the universe we know today. The distant galaxies are so far away that the light from them takes a long time to reach us.

The big bang model proposes that the universe started out unimaginably small, bright, hot, and dense, but has been expanding ever since (It now has a radius of about 15 billion light years). During the course of this expansion, some of the mass of the universe has condensed to form countless billions of galaxies of which there are about 10 billions in the known universe. These galaxies are grouped into clusters, which are themselves grouped into super clusters separated by vast distances in empty space. Astronomers can see faint galaxies 10000 million light years away. This means that the light from these galaxies has taken 10000 million years ago, and we do not know what these galaxies are really like today. As you look out into space, you are also looking back in time[4].

The expansion of the Universe is one of the most imposing discoveries of modern science. Today it is a firmly established concept and the only debate centers on the actual precipitating event that set our universe into being.

Quranic Viewpoint

The following verse of the Quran (sura 51, verse 47) where God is speaking may perhaps be compared with modern ideas:

-The heaven, We have built it with power. Verily, We are expanding it.

Chapter51: verse47

'Heaven' is the translation of the word 'sama' and this is exactly the extraterrestrial world that is meant. 'We are expanding it' is the translation of the plural present participle "musi'una" of the verb "ausa'a" meaning "to make wider, more spacious, to extend, to expand".

Some translators who were unable to grasp the meaning of the latter provide translations that appear to be mistaken, e.g. "we give generously" (R. Blachere). Others sense the meaning, but are afraid to commit themselves: Ramidullah in his translation of the Quran talks of the widening of the heavens and space, but he includes a question mark. Finally, there are those who arm themselves with authorized scientific opinion in their commentaries and give the meaning stated here. It refers to the expansion of the Universe in totally unambiguous terms[5].

This impressive statement in Holy Quran leads us to the scientific miracles of this supreme revelation.

References
1. Resurrection of Mankind through the Reversal of Time, by Dr. Muhammed Humayoun Khan, Illinois, USA.
2. Holy Quran translated by Marmaduke Pickthall, George Allen and Unwin Ltd., London, Fifth Edition, 1969.

3. Exploration of the Universe, by George Abell, Hot, Rinehart and Winston Inc., USA, 1969.
4. Space, Stars, Planets and Spacecraft, by Dorling Kindersley Publishers Limited, London, 1990.
5. The Bible, The Quran and Science (Le Bible, le Coran et la Science)," The Holy Scriptures Examined in the Light of Modern Knowledge, by Dr. Maurice Bucaille, French Physician, Seghers, Paris, 1987, English version published by North American Trust Publication, 1978.

Gaseous Origin of the Universe

ثُمَّ اسْتَوَىٰ إِلَى السَّمَاءِ وَهِيَ دُخَانٌ فَقَالَ لَهَا وَلِلْأَرْضِ ائْتِيَا طَوْعًا أَوْ كَرْهًا قَالَتَا أَتَيْنَا طَائِعِينَ ﴿١١﴾

The Glorious Quran states:

-Then turned He to the heaven when it was a smoke and said unto it and unto the earth: Come both of you, willingly or loathe. They said: We come obedient[1].

Chapter41: verse11

The essential point in this verse is that the origin of the universe is unambiguously considered to be smoke or something alike (gas). This is in compliance with the scientific discoveries revealed in recent history concerning this issue. To fully comprehend this concept, let us first deliberate the scientific viewpoint in this regard.

Scientific Viewpoint

The first hydrogen and helium atoms formed within minutes of the Big Bang explosion. These atoms eventually turned into clouds of gas, which around 14 billion years ago began to condense due to the effect of their own molecular gravity. Over the next 2 billion years, the first galaxies began to coalesce in these clouds of gas. Our own galaxy, the Milky Way Galaxy, began to come together around 10 billion years ago. Within this cloud of condensing gases, our own sun was born around 5 billion years ago, with the planets including earth, forming from the residual excess nebula of the spinning gas and dust[2].

The most famous theory about initial formation of the solar system is the theory of La Place. La Place postulated that: "Our sun was initially a huge and hot accumulation of flaming gases that was spinning, and because of this spinning motion, the centrifugal force, which was formed by this evolution, especially at the equator caused some heaps to be thrown away into space and each one (now called planets) located in an orbit and started to revolve around the sun. Because of this revolutionary motion, also, in turn, threw some heaps away which formed their moons."

Around 1900, two astronomers named Molton and Chamberlain presented a new theory, which was supported by other scientists. In their theory they stated: "We estimate that, billions of years ago, a giant planet passed nearby the sun. Because of its strong gravity force, a dreadful tide occurred on the surface of the sun. The strength of this tide was so enormous that some stacks were thrown away, and each of them formed a planet of the solar system".

Eventually, through debate and research, a combination of the two theories became most widely accepted by the scientists[3]. Most scientist ultimately disregarded La Place's concepts regarding the planets separation from the sun, but retained the general principles about the formation of the planets and sun from a huge heap of gas.

Quranic Viewpoint

The astronomers' observations on the nebulas and distant galaxies, which are being formed now, turn these theories to a scientific principle and prove that the existing universe was also initiated from a smoke-like heap of gas as what is first presented in the Holy Quran[4].

Professor Yoshihide Kozai remarks: "I say I am very much impressed by finding true astronomical facts in the Quran". Dr. Kozai is Professor of Emeritus at Tokyo University, Hongo, Tokyo, Japan, and was the Director of the National Astronomical Observatory, Mikata, Tokyo Japan.

Sheik Abdul Majid Zendani, a Professor of Islamic Studies in King Abdulaziz University in Jeddah, Saudi Arabia says about his meeting with Dr. Kozaki:

"We presented him a number of Quranic verses describing the beginning of creation and the heavens, and which deal with the relationship of the earth to the heavens. After studying these verses, Professor Kozai asked about the Quran and about the time when the Quran was revealed. He was informed that the Quran was revealed 1400 years ago, and then we asked him about the facts, which these verses contained. After each answer he was shown the Quranic text. He expressed his astonishment, saying that this Quran describes the universe as seen from the highest point; everything seen is distinct and clear. The one who said this, sees everything in existence. Seen from such a point, there is nothing, which can be unseen".

Professor Al-Zindani also stated: "We asked him whether at some point in time the firmament was in a form of smoke. He stated that all signs and indications are converging to prove that at one point in time the whole firmament was nothing but a cloud of *smoke*. This has come to be established as a proven visible fact. Scientists now can observe new stars forming up out of that smoke, which is the origin of our universe. The illuminating stars we see today were, just as was the whole universe, in the smoke form. We presented to him the Quranic verse saying:

-Then he turned to the sky, and it had been (as) smoke (dukhaan): He said to it and to the earth: come you together, willingly or unwillingly. They said: we do come (together) in willing obedience.
Chapter41: verse11

Some current scientists describe this dukhaan or smoke as "mist". But Professor Kozai pointed out that the term "mist" does not correspond to the description of this smoke, because mist is characteristically cold, whereas this cosmic smoke is somewhat hot. Dukhaan indeed is made up of diffused gases to which solid substances are attached, and this is the exact description of the smoke from which the universe emerged even before the stars were formed. Professor Kozai said that because that smoke was hot, we can not describe it as "mist". Dukhaan is the best descriptive word that can ever be. In this way Professor Kozai continued to scrutinize each Quranic verse we presented to him.

According to this same source, Dr. Kozai was finally asked: 'What do you think of this phenomenon which you have seen for yourself, namely, that science is beginning to discover the secrets of the universe, whereas many of these secrets have already been revealed in the Quran or in the Sunnah? Do you think that the Quran was given to the Prophet Muhammed (Peace Be Upon Him) from a human source?'

Professor Kozai replied: "I say, I am very much impressed by finding true astronomical facts in Quran, and for us modern astronomers have been studying a very small piece of the universe. We have concentrated our efforts for understanding of a very small part. Using telescopes, we can see only very few parts of the sky without thinking about the whole universe. So, by reading Quran and by answering to the questions, I think I can find my future way for investigation of the universe".

Professor Kozai believes it is impossible that the Quran was from a human source. He further stated that we scientists in our studies concentrated only on a small area, but if we read the Quran, then we will see a much larger picture of this universe. He contends that scientists have to look at it in a panorama, not within limited and narrow perspectives. Professor Kozai acknowledges relating to the Cosmos; he is now able to define his way in the future. He states that, from now on, he will plan his research guided by the comprehensive Quranic view of the universe[5].

This is the everlasting miracle, which renews itself. This is a miracle which gives life and which convinces Muslims and Non-Muslims, and which will convince all generations until the Day of Judgment. Allah tells all believing Muslims in Holy Quran:

-But Allah bears witness that what He has sent unto thee He has sent with His (own) knowledge.

Chapter4: verse166

-And say: Praise be to Allah, Who will soon show you His signs, so that you shall know them.

Chapter27: verse93

References
1. Holy Quran translated by Marmaduke Pickthall, George Allen and Unwin Ltd., London, Fifth Edition, 1969.
2. Eyewitness Encyclopedia of Space and the Universe, by Dorling Kindersley Publishers Limited, London, 1990.
3. The Quran and the Last Prophet (in Persian language), by ayatollah Makarem Shirazi, Dar AL-Kotob Al-Islamiah, Qum, the Islamic Republic of Iran, 1996.
4. ibid.
5. The Quran on the origin of the Universe: http://www.islam-guide.com/.

Extraterrestrial Life

وَمِنْ ءَايَٰتِهِۦ خَلْقُ ٱلسَّمَٰوَٰتِ وَٱلْأَرْضِ وَمَا بَثَّ فِيهِمَا مِن دَآبَّةٍ وَهُوَ عَلَىٰ جَمْعِهِمْ إِذَا يَشَآءُ قَدِيرٌ ۝

وَلَهُۥ مَن فِى ٱلسَّمَٰوَٰتِ وَٱلْأَرْضِ كُلٌّ لَّهُۥ قَٰنِتُونَ ۝

How likely is that there are other civilizations in our universe? Many scientists think that given the conditions and enough time, such as the intervening millenniums, the development of life is likely, if not inevitable.

Current scientific investigations state," life is not individualized to earth and the conditions for life are provided in many planets". It is mentioned in the book "Life in the World" published by Life Magazine Publications: "As the scientists have estimated, only in our Milky Way Galaxy may exist millions of stars whose subordinating planets are habitable".

Astonishingly enough, the Glorious and Holy Quran transcends the frontiers of time and speaks of extraterrestrial living creatures in the universe:

-*And of his portents is the creation of the heaven and the earth, and whatever beasts He hath dispersed therein, and He is able to gather them when He will*[1].

Chapter42: verse29

-Unto Him belongeth whosoever is in the heavens and in the earth. All are obedient unto Him[2].

Chapter30: verse26

The preceding verses along with some others explicitly indicate the existence of living creatures, which are dispersed in the heavens. To attain a vivid background of understanding, let us first glance over what science has to present in this regard. Then we will pay attention to the Quranic viewpoint.

Scientific Viewpoint

We have discussed that the organic molecules, namely, carbon_ based compounds that form the building blocks of life, as we know it. They are scattered abundantly throughout the galaxy. In interstellar clouds and newly fallen meteorites, astronomers have found complex organic molecules, including ethyl alcohol (the drinkable kind) and formaldehyde (embalming fluid). But a widespread abundance of these organic precursors does not, of course, guarantee that life is commonplace.

So far scientists and governments have not publicly found any trace of life elsewhere in the solar system. However, there are billions of stars are like the sun, and we know that one may even have planets similar to those in the solar system. Optical telescopes are not powerful enough to see another planet like the earth, but they can see the dust around stars where planets might form. SETI_ the Search for Extraterrestrial Intelligence_ is a group of scientists searching for intelligence life elsewhere in our galaxy.

SETI use radio dishes to collect radio signals from space and analyze them with a computer, which select signals that may have been produced artificially[3].

The initial step in the SETI effort, and the major development effort, involves design and construction of a spectrum analyzer, a radio receiver that can simultaneously search millions discrete channels in the microwave portion of the spectrum and then identify signals that might be of extraterrestrial origin.

Present theories of the origin and evolution of life suggest that life may not be unique to earth. On the contrary, theorists say, life may be widespread throughout the galaxy. Many scientists believe that, if life exists elsewhere, it could have evolved to a state of intelligence and curiosity and could have the ability to build the tools required for interstellar transmission and reception of radio signals.

If that is the case, the scientists believe other civilizations, too, could be searching for intelligent companions. Communication on a galactic scale could have evolved. So far, no signs of such signals have been detected or have been publicly acknowledged.

Pulsar Planets

Radio signals from the constellation Virgo led Penn State professor of Astronomy Alexander Wolszczan to discover the first planets ever known outside our solar system. He discovered the planets in 1991 and confirmed their existence in 1994.

Wolszczan discovered that radio signals were coming from a distant tiny star in the constellation Virgo, 7,000 trillion miles from Earth. Measurements he obtained from helped him to determine that two of the planets are similar in mass to Earth and the other is about the mass of the moon.

Until Professor Wolszczan's discovery, the only known planets were those in our solar system. His work suggests that planets may be more common in the universe than astronomers had previously thought. It also helped astronomers to understand how planets, including earth, were formed. The planets Wolszczan found may not support life because the tiny they orbit bombards them with deadly radiation. However, his discovery increases the chances that somewhere in the universe planets may exist that, like earth, are capable of supporting life.

Quranic Viewpoint

You sometimes hear people say, "There are millions of galaxies, and hundreds of billions of stars just in our own galaxy. Even if just a tiny fraction of the stars have planets, and even if just a tiny fraction of the planets have intelligent life, there must be many millions of other civilizations."

In recent years, astronomers have realized that the universe has to be incredibly finely balanced to make human life possible. The possibility of such a universe arising by blind chance seems incredibly small, nearly impossible. This is called the 'Anthropic Principle'.

Many other discoveries have seemed initially just as remote an impossible, but have been found to be true when great minds apply themselves and technology gives us the tools. (Such as the invention of the micro processing chip, a piece of silicon that holds libraries worth of information, Who would have thought this possible even 75-100 years ago?

To continue thought, we might discover definite evidence of alien civilizations tomorrow. Scientists like Carl Sagan and Paul Davies have argued that if this happened, it would somehow "disprove" all the world's religions. Is it not clear why this should Based upon the proof we are presenting here, it seems that Islam supports such a theory and eventual possibility that we are not alone!

Whatever the truth about alien civilizations, the fact that the universe is "just right" for us to live in is strong evidence that we are not alone - we are not here by chance. Someone has designed the universe for a purpose - to make our lives possible. The designer must be very powerful, and very intelligent.

Could it be that there is a God after all? And if there *is* a God, who has designed the whole universe to make human life possible, is it possible that He, the Glorified, is interested enough in us to want us to know what He is like? To communicate with us in a way that we can understand and respond to? That would really mean that we are not alone![4]

To continue, the word "Daabbah" (beast) in the first verse refers to any intelligent living creature (other than human being). In other words, the word "beasts" and the word "therein" explicitly remark: Living creatures are not specified to earth and the extraterrestrial intelligence exists somewhere in the universe.

The important note, which has to be necessarily stated, is that all of those who speak about life in other worlds presume that conditions for life should be similar to what is on earth; but this is maybe a reflection of man's narcissist thinking. The truth most probably is that life may appear in different forms, in the other worlds, which does not require the terrestrial conditions. For instance, if we had not ever seen the sea animals, we would have believed that nothing could live in the sea, because life depends on air, and there is no sensible air inside the water. But after studying about the sea animals' respiratory organs, it is found out that they also use oxygen, however, with a specific system, which enables them to respire and utilize the water's oxygen. Therefore, there may be specific living creatures whose special organism enables them to live in very high or very low temperatures and other conditions too[5].

The most Holy Quran, in the verses above, strongly clarifies the existence of living creatures in the worlds other than earth. This is the fact that scientists will undoubtedly conceive it in the future.

References
1. Holy Quran translated by Marmaduke Pickthall, George Allen and Unwin Ltd., London, Fifth Edition 1969.
2. ibid.
3. Exploration of the Universe, by George Abell, Hot, Rinehart and Winston Inc., USA, 1969.

4. The Universe in the classroom, Number 19, Astronomical Society of the Pacific, winter 1992.
5. The Quran and the Last Prophet (in Persian), by ayatollah Makarim Shirazi, Dar Al-Kotob Al-Islamiah publishers, Qum, the Islamic Republic of Iran, 1996.

Further Reading
Scientific Journals
- Wolszczan, A. and Frail, D., 1992, Nature, Vol. 255, pg. 145, "A Planetary System around the Millisecond Pulsar PSR1257+12".
- Wolszczan, A., 1994, Astrophysics and Space Science, Vol. 212, pg. 67, "Toward Planets Around Neutron Stars".
- Wolszczan, A., 1994, Science, Vol. 264, pg. 538, "Confirmation of Earth-Mass Planets Orbiting the Millisecond Pulsar PSR B1257+12".

Popular Literature
- Fienberg, R., 1992, Sky and Telescope, Vol. 83, No. 5, pg.493, "Pulsars, Planets, and Pathos".
- Folger, T., 1992, Discover, Vol. 13, No. 4, pg. 38, "Forbidden Planets".
- Kirshner, R., 1994, SCIENTIFIC AMERICAN, Vol. 271, No. 4, pg. 58, "The Earth's Elements".
- Mikolajewski, M. and Mikolajewski, J.,1992, Postepy Astronomii, Vol. 40, No. 2, pg. 53, "Kosmiczne zegary Aleksandra Wolszczana".
- Sagan, C., 1994, SCIENTIFIC AMERICAN, Vol. 271, No. 4, pg. 92 "The Search for Extraterrestrial Life".

Termination of the Universe

يَوْمَ نَطْوِي ٱلسَّمَآءَ كَطَيِّ ٱلسِّجِلِّ لِلْكُتُبِ كَمَا بَدَأْنَآ أَوَّلَ خَلْقٍ نُّعِيدُهُۥ وَعْدًا عَلَيْنَآ إِنَّا كُنَّا فَـٰعِلِينَ ۝

وَهُوَ ٱلَّذِى يَبْدَؤُا۟ ٱلْخَلْقَ ثُمَّ يُعِيدُهُۥ وَهُوَ أَهْوَنُ عَلَيْهِ وَلَهُ ٱلْمَثَلُ ٱلْأَعْلَىٰ فِى ٱلسَّمَـٰوَٰتِ وَٱلْأَرْضِ وَهُوَ ٱلْعَزِيزُ ٱلْحَكِيمُ ۝

ٱللَّهُ ٱلَّذِى رَفَعَ ٱلسَّمَـٰوَٰتِ بِغَيْرِ عَمَدٍ تَرَوْنَهَا ثُمَّ ٱسْتَوَىٰ عَلَى ٱلْعَرْشِ وَسَخَّرَ ٱلشَّمْسَ وَٱلْقَمَرَ كُلٌّ يَجْرِى لِأَجَلٍ مُّسَمًّى يُدَبِّرُ ٱلْأَمْرَ يُفَصِّلُ ٱلْـَٔايَـٰتِ لَعَلَّكُم بِلِقَآءِ رَبِّكُمْ تُوقِنُونَ ۝

فَٱرْتَقِبْ يَوْمَ تَأْتِى ٱلسَّمَآءُ بِدُخَانٍ مُّبِينٍ ۝

أَوَلَمْ يَرَوْا۟ أَنَّ ٱللَّهَ خَلَقَ ٱلسَّمَـٰوَٰتِ وَٱلْأَرْضَ بِٱلْحَقِّ إِن يَشَأْ يُذْهِبْكُمْ وَيَأْتِ بِخَلْقٍ جَدِيدٍ ۝

What will be the destiny of our universe? Is it really going to die one day? Scientists have set forth some theories on the termination of the universe. Does the Quran have anything to say about such a sophisticated subject and challenging question? To find the answer, the following verses should command our full attention:

-On the day when We shall roll up heaven as a scroll is rolled for the writings; as We originated the first creation, so We shall bring it back again, a promising binding on us, so We shall do[1].
Chapter21: verse104

-And it is He who originates creation, then brings it back again, and it is very easy for Him. His is loftiest likeness in the heavens and the earth; He is the All mighty, the All wise[2].
Chapter30: verse27

-Ö Each (Heavens) runneth unto an appointed termÖ[3]
Chapter13: verse2

-So be on the watch for a day when heaven shall bring a manifest smoke[4].
Chapter44: verse10

-If He will, He can put you away and bring a new creation; that is surely no great matter for God[5].
Chapter14: verse19 and 20

At first glance, by comparing the scientific and Quranic viewpoints, it is quite comprehensible that one day the universe as we know it will be brought back to its first state. According to the Big Bang theory, the universe was originated from a single, extremely small and dense matter, the primordial mass.

Having reviewed this theory previously, the final destination of our universe can be inferred from the first two verses. The final state deductively seems to be the initial state itself, which is what the scientists believe about the origin and the end of the universe. We can arrive at this conclusion that the universe will contract to its initial situation i.e. the primeval atom.

The fifth Quranic quote listed above, also, leads us to this idea that the universe might possibly start expanding again as a new creation, after being contracted to its first position. We present you an overview of the pertinent scientific theories and the respective Quranic viewpoint regarding this issue. To better apprehend the Quranic statements, the following scientific viewpoint can be helpful.

Scientific Viewpoint

Far far in the future, the expansion of the Universe will be the most important thing. The end of the Universe depends on how much dark matter there is between the galaxies, which is very hard to see. If there is not much of it then the Universe will keep expanding forever to end with a Big Emptiness. If there is a lot of dark matter, then the Universe will eventually stop expanding and start to shrink. If that happens then, very far in the future, the Universe will shrink back down to a point no bigger than an atom. It would be the Big Bang in reverse process. We have no way of telling what would happen after that, or even if "after that" has any meaning.

Astronomers have put forward three main theories about how the universe will end:

1) The first theory states that the universe will expand forever. This theory, known as the steady state theory holds that the universe has always existed and has set the same everywhere and at all times, never changing significantly. Under the steady state theory, new matter is always created to replace what we lose through the expanding universe. The steady state universe is infinitely large and old with no beginning or end[6].

According to the steady state theory, the overall condition of the universe looks the same from any viewpoint at any moment in time. It also suggests that the density of the universe remains constant, requiring the continuous creation of new material at a rate that exactly compensates for the expansion of the universe[7]. This is said to be achieved by the creation of new matter in the form of hydrogen gas, the raw material for the new stars. According to this theory, one hydrogen atom per liter of space is created. Of course, this expansion cannot be eternal, because the heavens move away from each other, so that they will eventually lose their energy. According to second law of thermodynamics, the entropy of whole universe is increasing. It means that at a moment, the heat energy will reach to a steady state throughout the universe and that is the day of absolute extinguishment of the stars. Life will be abolished and likely, the first situation will start. The heavens like the sea waves will return to the center of the sea after they had been moved towards the beach[8].

Another theory, which is called the oscillating universe theory, has also been recently put forward in this regard. It is called the oscillating universe theory, because it says that the universe expands and contracts over and over again:

ì If it expanded at once, then why can not it contract again? This theory says that it might just do that. The idea here is that the universe is expanding, but that the gravity of matter will eventually stop the growth, contracting it again until the next explosion. Time in the oscillating universe model begins the new with each big bangî.⁹

Quranic Viewpoint

Having noticed the scientific viewpoint, we can now realize the Glorious Quranís miraculous commentary on how the universe will end.

The word ìBadaínaî in Arabic meansî we originatedî or ìwe startedî. This simple but succinct word clearly states that the universe has an origin. So, the first theory, the steady state theory, which says that the universe has always existed with no beginning or end, is explicitly rejected by this verse.

In addition, the word ìNuíiduhuî means, ìWe will bring it backî. Subsequently, the second theory is also challenged by this verse, since the verse deems a return to the origin for the universe, which is seemingly contrary to this theory, stating that the universe will stop expanding and stabilizes when it reaches a certain size.

The miraculous Holy Quran strongly confirms the third theory which says:" *As We originated the first creation, so We shall bring it back again*". It means that the universe will eventually return to what had been in its initial moments of creation, which was extremely small, bright, hot, and dense mass. As mentioned before, the third theory is now the most acceptable theory among many scientists. This purport can also be achieved by the third verse saying:

-Ö *each (Heavens) runneth unto an appointed term*Ö

Chapter13: verse2

The forth verse also pictures the terminating universe as a smoky shape. The word ìDukhaanî in Arabic language means ìSmokeî. Interestingly enough, this word is used twice in the Quran. One is in describing the origin of universe, (see ìGaseous Origin of the Universeî) and the second one in describing the terminating universe in its finality. Therefore, from the viewpoint of the Quran, the ultimate state of the universe would turn to the same gaseous smoke as its origin. This is exactly what scientists now confirm:

ì According to the scientists, when a star dies, it turns to a supernova and then it explodes. Because of this explosion, its material disperses into space in the form of gaseous cloudsî ¹⁰. (See ìInterstellar Galactic Materialsî). Now, let us draw our attention to the fifth verse:

-*If He will, He can put you away and bring a new creation; that is surely no great matter for God.*

Chapter14: verse19 and 20

As we stated above, according to the oscillating theory, the universe might expand and contract repeatedly. Time in the oscillating universe model could begin the new with each big bang. The possibility of this theory could be easily understood from the verse. The word ì ifî implicitly indicates the conditional statement of this possibility, which hinges upon Godís will. Nonetheless, from the standpoint of science and the Holy Quran, it is at least quite likely, if not inevitable.

We can come to this conclusion that the universe will not expand forever and ultimately will be brought back to its first state of creation (or contracts) under the affection of the gravitational forces between all of the matters in the massive universe and it may be repeated again.

The Holy Quran is not a book of some prosaic words to be fixed in the ancient times. It comprises the best guidelines for the entire needs of Humankind. It not only reveals the tidings of the past, which no one had known them before it, but also transcends the frontiers of the present time.

The verses of the Holy Quran however, go far beyond modern cosmology in describing the future of the universe so that it really sounds to be a fresh and new book of all time. The miraculous Quran gives us insight into laws of the universe that are yet to be discovered in the future.

However, an important question is raised here. How could all this wonderful knowledge have been existed in a book since 1400 years ago? Could really the unlettered Muhammed whose job was watching the sheep in the desert of Arabian Peninsula, be the author of such an amazing book? In the sight of those who possess a right conscience, wisdom, and logic, Holy Quran is no less than a revelation from God Almighty.

Terminology

(*) Big Crunch: If there is enough matter in the Universe, eventually gravitational forces will stop its expansion. When this happens, gravity will cause the universe to reverse its direction and begin to collapse under its own weight. This phase of the Universe's life is known as the Big Crunch. Eventually all of the matter in the Universe will collapse into a super dense state and possibly even collapse into an unimaginably massive black hole. Some theorize that the Universe could collapse into the same singular state that it began as and then blow up in another Big Bang. In this way the Universe would last forever but would continually go through these phases of expansion and contraction, Big Bang and Big Crunch and so on...[11]

References

1. Holy Quran translated by Arthur J. Arberry, Ansarian Publishers, Qom, the Islamic Republic of Iran, 1999.
2. ibid.
3. Holy Quran translated by Marmaduke Pickthall, George Allen and Unwin Ltd., London, Fifth Edition 1969.
4. Holy Quran translated by Arthur J. Arberry, Ansarian Publishers, Qom, the Islamic Republic of Iran, 1999.
5. ibid.
6. Department of Education of the state of Maryland (Film Ideas).
7. Eyewitness Encyclopedia of Space and the Universe, by Dorling Kindersley Publishers Limited, London, 1990.
8. The Quran and the Last Prophet (in Persian language), by ayatollah Makarim Shirazi, Dar Al-Kotob Al-Islamiah publishers, Qom, the Islamic Republic of Iran, 1996.
9. Department of Education of the state of Maryland (Film Ideas).
10. The Birth and Death of Stars, by Isaac Asimov, Milwaukee, Gareth Stevens Publishing, 1989.
11. Windows to the Universe, at http://www.windows.ucar.edu/ at the University Corporation for Atmospheric Research (UCAR), The Regents of the University of Michigan.

Destiny of Earth

وَإِذَا ٱلْجِبَالُ سُيِّرَتْ ۝

وَإِذَا ٱلْبِحَارُ سُجِّرَتْ ۝

وَإِذَا ٱلْبِحَارُ فُجِّرَتْ ۝

وَتَكُونُ ٱلْجِبَالُ كَٱلْعِهْنِ ٱلْمَنفُوشِ ۝

وَإِذَا ٱلْجِبَالُ نُسِفَتْ ۝

According to leading scientists, the creation is physically mortal in all its aspects. The earth, like any other things in this universe is also bound to death. (See "Death of the Universe"). However, the earth's life depends on the energy, which receives from the sun. When the sun's energy ends up, the earth then is destined to die too.

What will our earth look like in its final day? Scientists have recently propounded theories, which represent the physics of earth in that situation. According to the astrophysicists, when the sun's energy ends up, it turns to a red giant and swells to a much bigger size than what it is now. Because of the sun's closeness to the earth, its extraordinary heat will melt the mountains and boils away all the seas on the earth.

The most Holy Quran has vividly clarified the pertinent physical situation of the dying earth through some verses as follows:

-When the mountains shall be set moving.
Chapter81: verse3

-When the seas shall be set boiling.
Chapter81: verse6

-When the seas swarm over.
Chapter82: verse3

-And the mountains shall be like plucked wool-tufts.
Chapter101: verse5

-When the mountains shall be scattered[1].
Chapter77: verse10

As vividly demonstrated by the preceding information, there is a firm concurrency between the scientific and Quranic viewpoints. To gain a strong proof for the authenticity of respective Quranic statements, the reference should be made to the latest scientific findings on this issue.

Scientific Viewpoint

According to most commonly held scientific predictions, the solar system will last for 4 to 5 billion years. As previously stated, the sun is going to lose its energy in the very distant future. It will run out of hydrogen to fuse and it will grow into a red giant star. The planets nearest the Sun (including the earth) will be burnt away. By that time some will have spread from Earth far out into the galaxy, or even to other galaxies[2].

What will happen to earth during this phase? When the Sun reaches this stage in its life, it will swell to 30-100 times its present size. Needless to say, if anything is still alive on the Earth, it will not survive this change. The Sun will become large enough to swallow Mercury, and bright enough to boil away the Earth's oceans and to melt the mountains.[3]

Quranic Viewpoint

Now let us revise the preceding verses of the Quran:

-When the mountains shall be set moving.

Chapter81: verse3

-When the seas shall be set boiling.

Chapter81: verse6

-When the seas swarm over.

Chapter82: verse3

-And the mountains shall be like plucked wool-tufts.

Chapter101: verse5

-When the mountains shall be scattered.

Chapter77: verse10

We do not think that any well-reasoned man has any problem with the authenticity of these verses, since no contradiction exists between science and the contexts of these Quranic statements indeed.

However, an important question becomes inescapable here. Have these statements really been made by the unlettered Muhammed (peace be upon him) who lived 1400 years ago in the Arabian Peninsula among the most awkward nations? Once again the most logical assumption is that Holy Quran could have not a human source, since the science did not have anything to say about the future of the universe in such a precise and authentic way in that time. The only explanation that lends itself to this is that the Quran must be a revelation from the creator of the whole universe, i.e. God Himself.

References
1. Holy Quran translated by Arthur J. Arberry, Ansarian Publishers, Qum, the Islamic Republic of Iran, 1999.
2. Descriptive Astronomy.
3. Penny Press Ltd. 2000.

Death of the Universe

يَوْمَ تَكُونُ ٱلسَّمَآءُ كَٱلْمُهْلِ ۝

فَٱرْتَقِبْ يَوْمَ تَأْتِى ٱلسَّمَآءُ بِدُخَانٍ مُّبِينٍ ۝

فَإِذَا ٱنشَقَّتِ ٱلسَّمَآءُ فَكَانَتْ وَرْدَةً كَٱلدِّهَانِ ۝

As being discussed before, according to scientists, the life of our universe is destined to terminate one day. It will eventually contract to reach its initial position, due to the lack of enough energy throughout the universe to continue expanding.

The Holy Quran presents many verses concerning the events of the Day of Resurrection and describes the individual physical characteristics of these dreadful events to provide us the insight into this eventuality and assure the humankind of man's individually accountability before God Almighty. Among these verses, some clearly describe the physics of the Day of Judgment:

-*The day when the sky will become as molten copper[1].*

Chapter70: verse8

-*So be on the watch for a day when heaven shall bring a manifest smoke[2].*

Chapter44: verse10

-And when the heaven splitteth asunder and becometh rosy-like red hide[3].

Chapter55: verse37

Before contemplating the verses, we will first go over the respective scientific illustrations. To achieve a better image of what the Holy Quran dictates in this regard, we have provided you the following detailed scientific passage.

Scientific Viewpoint

Death of a star

Stars evolving and dying? These may be strange concepts to many readers, since most of us think of stars as symbols of permanence. Compared to a human lifetime, stars appear eternal. But the general consensus of modern day scientists theorized that in the course of millions to billions of years, they do evolve and age, and eventually they die. The most massive stars end their lives violently and spectacularly. Less massive stars die more peacefully. Examples of the former are supernova explosions, which are among the most energetic events in the heavens. Examples of the latter are the formation of planetary nebulae, in which, over the course of thousands of years, a star sheds its outer layers.

After stars have exhausted their hydrogen fuel, they are unable to support the weight of the overlying layers with their internal gas pressure. This crisis means that they must renew the battle against gravitational compression through a sequence of contracting, heating, and igniting new sources of nuclear fuel in order to survive.

During most of a star's lifetime, nuclear fusion in the core generates enough outward pressure to exactly balance the inward pull of gravity associated with the star's mass. As the nuclear fuel is exhausted, the outward forces diminish, allowing the gravitation to compress the star inward. Eventually, all possible nuclear fuel is used up and the core collapses.

How far and fast the star's collapses is determined by the star's initial mass. If the star is massive enough, it may explode and then collapse into a supernova. If it is less massive, it may become a white dwarf.

Death of an Ordinary Star

After a low-mass star like the Sun exhausts the supply of hydrogen in its core, there is no longer any source of heat to support the core against gravity. Once again, as in the case discussed above the core of the star collapse under gravity's pull until it reaches a high enough density to start burning helium to carbon. Meanwhile, the stars' outer envelope expands and the star evolves into a red giant.

Our sun will eventually evolve into a red supergiant as it exhausts the helium in its core. At this stage, it will have an outer envelope extending out towards Jupiter. During this brief phase of its existence, which last only a few tens of thousands of years, the Sun will lose mass in a powerful wind. Eventually, the Sun will lose all of the mass in its envelope and leave behind a hot core of carbon imbedded in a nebula of expelled gas. Radiation from this hot core will ionize the nebula produces a striking "planetary nebula", much like the nebulas seen around the remnants of other stars. The carbon core will eventually cool and become a white dwarf, the dense dim remnant of a once bright star.

Death of a Massive Star

Massive stars burn brighter and perish more dramatically than most. When a star ten times more massive than sun exhausts the helium in the core, the nuclear burning cycle continues. The carbon core contracts further and reaches high enough temperature to burn carbon to oxygen, neon, silicon, sulfur and finally to iron. Iron is the most stable form of nuclear matter and there is no energy to be gained by burning it to any heavier element. Without any source of heat to balance the gravity, the iron core collapses until it reaches nuclear densities. This high density core resists further collapse causing the falling matter to "bounce" off the core. This sudden core bounce (which includes the release of energetic neutrinos from the core) produces a supernova explosion. For one brilliant month, a single star burns brighter than a whole galaxy of a billion stars. Supernova explosions inject carbon, oxygen, silicon and other heavy elements up to iron into interstellar space. They are also the site where most of the elements heavier than iron are produced. This heavy element enriched gas will be incorporated into future generations of stars and planets. Without supernova, the fiery death of massive stars, there would be no carbon, oxygen or other elements that make life possible.

White Dwarfs

For some time, the surface of the star may be as hot as several hundred thousand degrees Fahrenheit. In the course of a few hundred million years, the star cools and becomes a white dwarf. "White" refers to the color of the light it emits; "dwarf" refers to its small size.

White dwarfs give scientists insight into a type of matter that cannot be studied in terrestrial laboratories. They also show us what the Sun will eventually look like, with direct implications for future conditions on Earth.

All medium-sized stars like the Sun eventually collapse to form white dwarfs. It has been estimated that there are at least a billion white dwarfs in our galaxy.

Mathematical modeling of white dwarfs leads to a surprising result: The larger its mass is, the smaller will be its radius. For example, a white dwarf whose mass is 0.4M, has a radius equal to about 1.5 percent of the Sun's radius, or about 10,000 km. But a white dwarf whose mass is 0.8M, has a radius that is about 1 percent of the Sun's radius, or about 7000 km (about the size of Earth). An object the size of an olive made of this material would have the same mass as an automobile!

Red Giant

Scientists predict that our own Sun will eventually become a Red Giant one day a very long time from now (a few billion years). When this happens it will swell up and get so large that the orbit of the earth will actually be inside the sun! Since the sun will still weigh about as much in the future as it does now, but it will be taking up much more space, its weight will have to be very spread out. A spoon-full of material taken from a red giant (except at its very center) will not be heavy at all, weighing much less than a small grain of salt[4].

The Sun's basic energy source is hydrogen, a non-renewable resource. A series of nuclear fusion reactions convert hydrogen nuclei into helium nuclei in the central parts of the sun and give off energy in the process in the form of light and heat. When this hydrogen is used up, the core of the Sun will collapse. A shell of hydrogen on the edge of the collapsed core will be compressed and heated. So in essence the very light and heat that makes life possible here on earth is not limitless and will eventually cease to be one day.

Scientists theorize that towards the end of any star's life, the temperature near the core rises and this causes the size of the star to expand. This is the fate of the sun in about 5 billion years. You might want to mark your calendar!

The hydrogen is depleted so it no longer generates enough energy and pressure to support the outer layers of the star. As the star collapses, the pressure and temperature rise until it is high enough for helium to fuse into carbon, i.e. helium burning begins. To radiate the energy produced by the helium burning, the star expands into a Red Giant[5].

Meanwhile inside the shell, the core of the star shrinks and heats up enough to fuse the helium nuclei together into even heavier ones. Among the commonest nuclei are carbon, nitrogen and oxygen. Heavier and heavier nuclei are created inside a red giant, the heaviest nearest the middle. Iron is a type of atom, which was created in red giant stars.

Supergiant Stars

Supergiant stars differ from giant stars in much the same manner that giant stars differ from main sequence stars. They have depleted their helium core fuel, and go on to burn carbon. They continue to burn helium and hydrogen in shells around the core, but carbon is being burned in the core at this stage. An important thing to remember is that each time a new and heavier element is being burned in the core, the time that this takes place in is much shorter than the previous time period. For example, a 15 solar mass star will burn hydrogen for about 10 million years, helium for 1 million years, carbon for 300 years, oxygen for 200 days, and silicon for 2 days. No more fusion takes place after this point. Also, the more massive a star is, the faster it burns out its fuel. The most stable element (and the last to be produced via silicon burning) is Iron (Fe) 56.

At this point in a star's life, it starts losing mass. Some of the lighter elements start getting blown away by the star. This mass loss continues at different times throughout the star's life.

Supernovae

Scientists have theorized that if the mass of the white dwarf becomes greater than about 1.4 times the mass of the sun-called the Chandrasekhar limit-it will collapse. In a binary star system this could happen if a nearby companion star dumps enough material onto a white dwarf to push it over the Chandrasekhar limit. The resulting collapse and explosion of the white dwarf is believed to be responsible for the so-called Type Ia supernovas. When a star explodes it can shine with the brilliance of a billion suns.

Supernovae are the creative flashes that renew the galaxy. They seed the interstellar gas with heavy elements, heat it with the energy of their radiation, stir it up with the force of their blast waves and cause new stars to form.

When a massive star has used up its nuclear fuel and the pressure drops in the central core of the star, the matter there is crushed to higher and higher densities and temperatures of billions of degrees. Under these extreme conditions, nuclear reactions occur violently and catastrophically reversing the collapse. A thermonuclear shock wave races through the now expanding stellar debris, fusing lighter elements into heavier ones and producing a brilliant visual outburst that can be as intense as the light of ten billion suns!

For massive stars, once silicon burning commences, their end is near for these stars have exhausted all available nuclear fuels that have sustained them throughout their lives against the compressional forces of gravity. We believe that the result of nuclear fuel exhaustion is a supernova outburst with the principal question being what is the actual physical mechanism for this catastrophe?

The derived temperature for the supernova outburst is about 50 billion Kelvins [6]. A temperature that is perhaps beyond the comprehension of man. The facts demonstrate that the supernova--a single star--can be seen as a naked-eye object, outshining all those billions of suns, shows the great power of a supernova explosion.

In one second, 170,000 years ago, the exploding star released more energy than our Sun will during its entire ten billion-year lifetime. By comparison, the Sun puts out the same amount of energy as is contained in all the world's nuclear weapons in only ten millionths of a second.

Quranic Viewpoint

We presented you the scientific viewpoints regarding the death of stars, which accordingly lead us to the ultimate physical situation of the terminating universe.

These verses should attract our full attention:

-The day when the sky will become as molten copper,

Chapter70: verse8

-So be on the watch for a day when heaven shall bring a manifest smoke.

Chapter44: verse10

-And when the heaven splitteth asunder and becometh rosy like red hide.

Chapter55: verse37

The first verse describes the sky in the Day of Judgment as "Molten Copper". From the standpoint of science, this description of the death of the universe is deemed to be a wonderful similitude. As you read in "Scientific Viewpoint", when a star dies, it converts to a white dwarf, red giant, red supergiant and eventually a supernova. In all these stages of dying procedure, the carbon core contracts further and reaches high enough temperature to burn carbon to heavy elements like oxygen, neon, silicon, sulfur and finally to heavier metals like iron.

This is how scientists illustrate the destiny of a dying star and its final materials:

"The core of the original star is made of iron. During and after the explosion, the iron and other elements in the star's outer layers are fused together by the tremendous heat and energy to form new elements, in a process called "nucleosynthesis." In fact, almost all atoms heavier than iron are thought to be created in supernova explosions. When the star explodes, the new materials are blasted out into empty space, eventually to condense and be incorporated into new stars and planets. Many of the atoms that make up our own Earth and the life that inhabits it therefore originated in the cosmic furnace of some distant supernova".[7]

According to the Arabic encyclopedias, the word "Muhl" in the verse means "iron, molten copper, brass and generally metal"[8]. So, all the sky becomes extremely hot and dark red gas like a molten copper (metal), which is miraculously illustrated in this verse. Astonishingly enough, this descriptive commentary is accordingly in compliance with the scientific facts.

The word "Dukhaan" in Arabic language means "Smoke". (See "Gaseous Origin of the Universe). The second verse also describes the terminating universe as a "manifest smoke". As we formerly mentioned, when a star dies, it converts to a white dwarf, red giant and finally supernovae. In that situation, stars are put out and eventually explode. Because of this explosion, the materials disperse into space in the form of gaseous nebulas and clouds[9]. (See "Interstellar Gallactic Materials"). This is also what scientists state apropos the death of stars: "When a white dwarf forms, however, a lot of gas is blown off (but not in a supernova explosion) to form a "planetary nebula," which looks like a giant smoke ring. The still-glowing core that is left over is a white dwarf. When it cools and no longer glows, it is a black dwarf".[10]

Therefore, this is also considered to be an amazing description of the situation of a dying universe after explosion, which is in accordance with the scientific statements.

A Dazzling Evidence

-And when the heaven splitteth asunder and becometh rosy-like red hide.

Chapter55: verse37

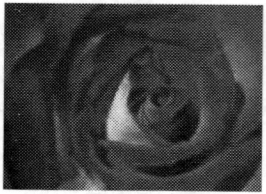

Comparison: Above, the real red rose. Below, the cat's eye nebula seems like a red rose. A nebula (Gaseous Cloud) is the remainder of a dying star. Wonderfully, this is what Holy Quran prophesied for the future of our universe.

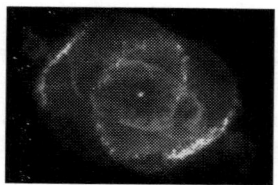

The Cat's Eye Nebula
Credit: J.P. Harrington and K.J. Borkowski (U. Maryland HST NASA) October 31, 1999

-And when the heaven splitteth asunder and becometh rosy-like red hide.

Chapter55: verse37

NASA's Explanation

NASA states that "Three thousand light-years away, a dying star throws off shells of glowing gas. This image from the Hubble Space Telescope reveals the Cat's Eye Nebula to be one of the most complex planetary nebulae known. In fact, the features seen in the Cat's Eye are so complex that astronomers suspect the bright central object may actually be a binary star system".

It is mentioned that this represents a dying star, which is inevitably the awaited fate of our own solar system, which the Holy Quran speaks of as the day when the Sky is torn apart and becomes rosy-like red hide (As seen in the comparative images).

Again this question is raised here: How could all these wonderful knowledge have existed in a book 1400 years ago? For those who possess right minds and open hearts; for those who are seeking the truth; for those who reflect in the creation of the universe and for those who own a pure logic, Holy Quran is no less than a revelation from God Almighty. It is indeed a favor for us to contemplate and dwell upon.

References
1. Holy Quran translated by Marmaduke Pickthall, George Allen and Unwin Ltd., London, Fifth Edition 1969.
2. Holy Quran translated by Arthur J. Arberry, Ansarian Publication, Qum, the Islamic Republic of Iran, 1993.
3. Holy Quran translated by Marmaduke Pickthall, George Allen and Unwin Ltd., London, Fifth Edition 1969.
4. Neutron Stars, White Dwarfs and Red Giants, http://van.hep.uiuc.edu/ by Marta Lewandowska, July 1999 CWRU Astronomy Dept., Case Western Reserve University.
5. NASA Educational Briefs, "The Death of a Star: Supernova 1987a," EB-88-1 (S).
6. Physics and Astronomy Department, George Mason University, Maintained by J. C. Evans.
7. By Jeff Silvis, Imagine the Universe, A service of the High-Energy Astrophysics Science Archive Research Center (HEASARC) http://heasarc.gsfc.nasa.gov/within the Laboratory for High Energy Astrophysics at NASA's Goddard Space Flight Center, http://heawww.gsfc.nasa.gov/.
8. Encyclopedia of Quran (Qaamousse Quran), by Ali Akbar Qurashi, Dar Al kotob Al-Islamiah, Teheran, Islamic republic of Iran.
See Also: Monjid Al-Tollab (Arabic-Arabic Encyclopedia), by Fu'aad Afraum Al-Bostaani.
9. The Birth and Death of Stars, by Isaac Asimov, Milwaukee, Gareth Stevens Publishing, 1989.
10. http://www.antwrp.gsfc.nasa.gov/.

Astronomy in Holy Quran

- Interstellar Galactic Material
- Revolution of the Heavens
- Black Holes
- Pulsars and Neutron Stars
- Sun and Moon and their orbits
- Inherent Diversity between Sun and Moon
- Revolution of Sun
- Composition of Meteors
- Motion of Earth
- Trundling Earth
- Spherical Earth

Interstellar Galactic Material

لَهُ مَا فِى ٱلسَّمَوَٰتِ وَمَا فِى ٱلْأَرْضِ وَمَا بَيْنَهُمَا وَمَا تَحْتَ ٱلثَّرَىٰ ۝

Scientists tell us that as a primary nebula condenses, it further divides into fragments. These fragments or galactic masses further split up into stars and their sub-products, the planets. Each time such a division or split occurred, there remains extra material apart from the principal elements newly formed. The scientific name for these extra materials is 'interstellar galactic material'.

Is this extra material or interstellar galactic material significant? Experts in astrophysics are quite aware of such material, which may have "a tendency to interfere with photometric measurements". The extra material is so rare that they are sometimes referred to as dusts or smokes or gases. Yet they altogether occupy so much total space that they may correspond to" a mass possibly greater than the total mass of the galaxies".

The Holy Quran mentions a rather curious category of created things, namely things between the heavens and the earth. Dr. Maurice Bucaille, the famous French Physician and the author of the book "The Bible, The Quran and Science (Le Bible, le Coran et la Science)", remarked that the interstellar galactic material is in fact mentioned in Holy Quran and "may surprise the twentieth century reader of the Quran". For example, one verse says as follows:

-To Him (God) belongs what is in the heavens, on earth, between them and beneath the soil[1].

Chapter20: verse6

What is that between the heavens and the earth? Dr. Bucaille explains as follows: "The creation outside the heavens and outside the earth is a priori (concept) difficult to imagine. To understand these verses, reference must be made to the most recent observations on the existence of cosmic extragalactic material, and one must indeed go back to ideas established by contemporary science on the formation of the universe...".

Scientific Viewpoint

Empty Space Is Not Empty

The Milky Way Galaxy is populated with dusty clouds of turbulent gas, exploding stars and evolving star systems, and the empty vacuum of space in between filled with a mixture of hydrogen and other gases commonly called the interstellar medium. The rarefied interstellar medium, abbreviated ISM, contains less than one particle per cubic centimeter (about the size of a thimble), compared to approximately 10 billion billion (a 10 with 18 zeros after it!) particles per cubic centimeters in the atmosphere here on Earth. Astronomers are eager to learn about the density, temperature and chemical composition of the ISM in order to better understand how stars are born and how they die[2].

Interstellar Medium

Interstellar medium is the region between the stars that contains vast, diffuse clouds of gases and minute solid particles. Such tenuous matter in the interstellar medium of the Milky Way system, in which the Earth is located, accounts for about 5 percent of the Galaxy's total mass. The interstellar medium is filled primarily with hydrogen gas. A relatively significant amount of helium has also been detected, along with smaller percentages of such substances as calcium, sodium, water, ammonia, and formaldehyde. Sizable quantities of dust particles of uncertain composition are present as well. In addition, primary cosmic rays travel through interstellar space, and magnetic fields thread their way across much of the region.

In most cases, interstellar matter occurs in cloud-like concentrations, which sometimes condense enough to form stars. These stars, in turn, continually lose mass, in some instances through small eruptions and in others in catastrophic explosions known as supernovae. The mass is thus fed back to the interstellar medium, where it mixes with matter that has not yet formed stars. This circulation of interstellar matter through stars determines to a large degree the amount of heavier elements in the cosmic clouds. Interstellar matter in the Milky Way Galaxy is found primarily in the system's outer parts (*i.e.,* the so-called spiral arms), which also contain a large number of young stars and nebulae. This matter is closely concentrated in a plane, a flat region commonly known as the galactic disk. The interstellar medium is studied by several methods.

Until the mid-20th century, virtually all information was obtained by analyzing the effects of interstellar matter on the light from distant stars with the aid of optical telescopes. Since the early 1950s, much research has been conducted with radio telescopes, which enable astronomers to study and interpret radio waves emitted by various constituents of the interstellar medium. For example, neutral (*i.e.,* non-ionized) hydrogen atoms absorb or emit very small amounts of radio energy of a particular wavelength--namely, 21 cm. By being measured at this point and compared with nearby wavelengths, absorbing or radiating hydrogen clouds can be detected. Optical and radio emissions have provided much of the information on the interstellar medium. In recent years, the use of infrared telescopes on orbiting satellite observatories has also contributed to knowledge of its properties, particularly the relative abundances of the constituent elements.

Latin: "mist" or "cloud", plural nebulae, or nebulas, any of the various tenuous clouds of gas and dust that occur in interstellar space. The term was formerly applied to any object outside the solar system that had a diffuse appearance rather than a point-like image, as in the case of a star. This definition, adopted at a time when very distant objects could not be resolved into great detail, unfortunately includes two unrelated classes of objects: the extragalactic nebulae, now called galaxies, which are enormous collections of stars and gas; and the galactic nebulae, which are composed of the interstellar medium (the gas between the stars, with its accompanying small solid particles) within a single galaxy. Today the term nebula generally refers exclusively to the interstellar medium.

The most conspicuous property of interstellar gas is its clumpy distribution on all size scales observed, from the size of the entire Milky Way Galaxy (about 1020 meters, or hundreds of thousands of light-years) down to the distance from the Earth to the Sun (less than 1011 meters, or a few light-minutes). The large-scale variations are seen by direct observation; the smallest are observed by fluctuations in the intensity of radio waves, similar to the "twinkling" of starlight caused by unsteadiness in the Earth's atmosphere. Various regions exhibit an enormous range of densities and temperatures.

Quranic Viewpoint

Now, with this scientific background, we may take again, the following verses into a deep consideration:

-To Him (God) belongs what is in the heavens, on earth, between them and beneath the soil.
Chapter20: verse6

-(Allah) Who created the heavens and the earth, and what between them is, in six days.
Chapter25: verse59

-We have not created the heavens and the earth, and what between them is, save with the truth and a stated term; but the unbelievers are turning away from that they were warned of.
Chapter46: verse3

Needless to say, all these verses implicitly point out to the existence of "interstellar galactic material" in the universe.

Again, we must face up to the implication of all this. How could a man living fourteen hundred years ago have known about interstellar galactic material? Was Mohammed, on whom be peace, well-versed in modern astrophysics? Or is the Quran nothing but the Very Word of God? Allah, the only true God, declares in His book: "The revelation of the scripture is from Allah, the Mighty, the Wise[3]:

-Surely We [Allah] have revealed the scripture unto you [Muhammed] with truth; so worship Allah, making religion pure for Him (only).
Chapter39: verse1 and 2

References
1. Holy Quran translated by Abdullah Yusuf Ali, Dar Al Arabia Publication, Beirut, Lebanon, 1968.
2. ORFEUS Science, http://www.snoopy.gsfc.nasa.com/.
3. The Bible, The Quran and Science (Le Bible, le Coran et la Science)," The Holy Scriptures Examined in the Light of Modern Knowledge, by Dr. Maurice Bucaille, French Physician, Seghers, Paris, 1987, English version published by North American Trust Publication, 1978.

Revolution of the Heavens

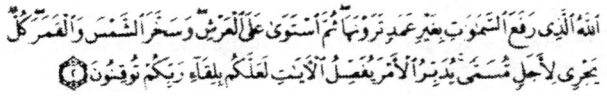

Today, it is a well-known fact that all the celestial bodies, without exception, are in motion, and the solar system is inside a galaxy, which is in motion as well. (See "Expanding Universe"). However, fourteen hundred years ago, the Holy Quran revealed to believing men and women that all heavens are in motion unto an appointed term:

-Allah, it is who raised up the heavens without visible supports, then mounted the throne, and compelled the sun and the moon to be of service, each runneth unto an appointed term; He ordereth the course, He detaileth the revelations, that haply ye may be certain of the meeting with your lord[1].

Chapter13: verse2

In this verse, there is a term, which enhances the majesty of Holy Quran and clarifies its scientific aspect of preternaturalism; i.e. the word "each runneth". The deleted pronoun refers to the moon, the sun, and also to all heavens.

Scientific Viewpoint

As we mentioned before, Over 400 years ago, Copernicus advanced a heliocentric-Sun Centered-theory. He suggested that the retrograde motion of the planets could be readily explained if the sun, rather than the earth, was at the center of the universe; that the earth is a planet; and that the planets move around the sun in circles.

Copernicus' theory although it put the sun instead of the earth at the center of the solar system (and subsequently the universe) still assumed that the orbits of celestial bodies had to follow such perfect orbits. This illustrates that Copernicus had not been able to break away entirely from the old ideas. The so called Copernican revolution, in the sense that our modern approach to science involves a comparison of nature and understanding really occurred later [2].

Keeping in mind modern concepts on the formation of the Universe reference was made to the evolution that took place, starting with primary nebula through to the formation of galaxies, stars and (for the solar system). This continues on to the appearance of planets beginning with the Sun at a certain stage of its evolution. Modern data lead us to believe that in the solar system, and more generally in the Universe itself, this evolution is still continuing.

In addition to our galaxy's rotation, it moves outward the assumed center of the universe. According to its expansion, all the galaxies, nebulas, stars and planets are all moving away from each other; in other word, from the assumed center of the universe in an incredible and astronomical speeds [3].

Quranic Viewpoint

How can anybody who is cognizant of these ideas fail to observe that with certain statements found in the Quran the manifestations of divine omnipotence are referred to? The Quran reminds us several times that: "(God) subjected the sun and the moon: each one runs its course to an appointed term."

This sentence is to be found in chapter 13, verse 2; Chapter 31, verse 29; chapter 35, verse 13 and chapter 39, verse 5. In addition to this, the idea of a settled place is in associate with the concept of a destination place in chapter 36, verse 38: "The Sun runs its course to a settled place. This is the decree of the All Mighty, the Full of Knowledge."

Settled place' is the translation of the word "mustaqarr" and there can be no doubt that the idea of an exact place is attached to it. (See "Revolution of Sun").

How do these statements fare when compared with data established by modern science?[4] In the time of revelation of Quran and even 1000 years later when the earth or sun were supposed to be stationed at the center of the universe these enlightening verses explicitly revealed the general motion of all heavens.

References
1. Holy Quran translated by Marmaduke Pickthall, George Allen and Unwin Ltd., London, Fifth Edition 1969.
2. Exploration of the Universe, by George Abell, Hot, Rinehart and Winston Inc., USA, 1969.
3. The Quran and the Last Prophet (in Persian language), by ayatollah Makarem Shirazi, Dar AL-Kotob Al-Islamiah, Qum, the Islamic Republic of Iran, 1996.
4. The Bible, The Quran and Science (Le Bible, le Coran et la Science)," The Holy Scriptures Examined in the Light of Modern Knowledge, by Dr. Maurice Bucaille, French Physician, Seghers, Paris, 1987, English version published by North American Trust Publication, 1978.

Black Holes

فَلَا أُقْسِمُ بِمَوَاقِعِ ٱلنُّجُومِ ۝

وَإِذَا ٱلنُّجُومُ ٱنكَدَرَتْ ۝

فَإِذَا ٱلنُّجُومُ طُمِسَتْ ۝

وَمَآ أَدْرَىٰكَ مَا ٱلطَّارِقُ ۝ ٱلنَّجْمُ ٱلثَّاقِبُ ۝

تَعْرُجُ ٱلْمَلَـٰٓئِكَةُ وَٱلرُّوحُ إِلَيْهِ فِى يَوْمٍ كَانَ مِقْدَارُهُۥ خَمْسِينَ أَلْفَ سَنَةٍ ۝

Black Holes are the ultimate demise of many stars. Undoubtedly, they are the most mysterious celestial bodies found in the sky. To say that the idea of a Black Hole is unfamiliar is a very mild way of putting matters. Their physical features are so strange that even the most well-versed would be pressed to describe them in every day language. We have to try to visualize something, which is completely outside our normal experience, and accept a situation in which we have to abandon not only the accepted laws of nature but also what we usually call common- sense. They could be viewed as a distinct exercise in faith.

Black Holes were first noted in the works of John Wheeler in 1969. A black hole is a theorized celestial body whose surface gravity is so strong that no light can escape from it. So we can not see them even with most powerful telescopes. Their light is extinguished or rubbed out.

The existence of black holes and their physical conditions are explicitly mentioned in the Quran. The following verses can lead us to a dazzling topic; Black Holes in the Quran:

-*When the stars become dim*[1].

Chapter77: verse8

-*And when the stars darken*[2].

Chapter81: verse2

-*No! I swear by the fallings of the stars*[3].

Chapter56: verse75

-*And what shall teach thee what is the night-star? The piercing star!*[4]

Chapter86: verse2 and 3

- *(Whereby) the angels and the spirit ascend unto Him in a Day whereof the span is fifty thousand years*[5].

Chapter70: verse4

Black holes are the dead stars into which matter has *fallen* and have become *dim* and *darkened*. They *pierce* the space around them and the time passes more and more slowly in them than passes on earth. Much greater intervals of earth-time, maybe *thousands of years*, would pass on earth comparing to what might pass in course of a year in the black hole. It is as if time is slowing down, or has even stopped.

Although, summarizing any such issue into a few pages is bound to lead to omission and over simplification, however, we attempt to present you the wonderful concurrency between the Quranic statements and the pertinent scientific viewpoints.

Scientific Viewpoint

Black Hole's Formation

There are two main processes constantly going on in massive stars: nuclear fusion, which tends to blow the star's hydrogen outward from the star's center and gravitation, which tends to pull all hydrogen back in the direction it had come. These two processes balance one another until all the star's hydrogen is exhausted, allowing gravitation to take over. Once gravitation dominates, the star becomes unstable and starts to collapse. Once the star starts to collapse, it does not stop, and the star (and ultimately its atoms) will cave inward upon itself, resulting in the formation of a black hole.

However, not all stars upon gravitational collapse form black holes. A star less than 1.4 times the mass of the sun will become a white dwarf. A star between 1.4 and 3 times the mass of the sun will become a neutron star. It is only those stars greater than 3 times the mass of the sun that become black holes upon collapse.

Additionally, a black hole can be formed by compression through external forces. This type of black hole is called a primordial black hole[6].

Once nuclear fusion is no longer possible, the neutron star has no new source of internal energy generation. With time, its rotation should slow and its magnetic field should decrease. Unless the neutron is "spun-up", it will eventually become *"invisible"*[7].

Some scientists have suggested that white dwarfs may be the mysterious dark matter in our galaxy. White dwarfs are burned-out stellar cinders. They cool and dim, and eventually, after some billions of years, become virtually undetectable, even if they reside in our galactic neighborhood. Because they are hidden from our view, there may be a vast number of unknown entities beyond our field of vision. If this proves to be true, white dwarfs will be the most common form of matter in our galaxy and possibly in the universe.

Black Holes, Gravity's Ultimate Victory

When the core of a neutron star becomes so small that the escape velocity at its surface exceeds the speed of light, it becomes a black hole.

A black hole has no "surface" but does have a size scale associated with it. Sometimes called the event horizon, this describes a sphere around the black hole separating objects which can (in theory) escape the black hole's gravity from those which cannot. At the event horizon, the escape velocity is the speed of light; inside the event horizon, not even light can escape the black hole[8].

The event horizon is the gravity field of a black hole where the space-time is so bent that light cannot escape it. The event horizon creates a region in space where nothing can escape, if nothing can go beyond the speed of light. Thus when something enters the event horizon, it will vanish without a trace. Should the object be emitting something, after it is enveloped by the event horizon, not even the emissions that traced its existence will escape the black hole.

The event horizon of a black hole grows larger as it accumulates more mass. In the centers of many galaxies, we believe there are black holes that have accumulated so much mass (from eating interstellar gas and stars). These supermassive black holes may have quite different origins from the stellar black holes that result from individual massive stars.

For a more in-depth understanding of black holes, we need to be familiarized with some of the basic ideas of general relativity, including the curvature of space and the gravitational redshift.

Curved Space-time

The second fundamental principle of General Relativity is that the presence of matter curves space. In this view, gravity is not a force, as described by Newton, but a curvature in the fabric of space, and objects respond to gravity by following the curvature of space in the vicinity of a massive object. The description of the curvature of space is the mathematically complicated part of general relativity involving "metrics", which describe the way that matter curves space, and tensor calculus.

Einstein's view is that the planets follow the curvature of space around the sun (and produce a tiny amount of curvature themselves)[9].

Quranic Viewpoint

Dark Stars

As we stated before, (see "Termination of the Universe"), scientists now believe that the heavens move away from each other, so that they will eventually lose their energy. According to second law of thermodynamics, the entropy of whole universe is continuously increasing. It means that at a moment, the heat energy will reach to a steady state throughout the universe and that is the day of absolute putout of the stars[10]. This is thought to be the ultimate situation of all stars to become a black hole, white dwarf, neutron star etc, depending on their mass. As discussed before, all of them eventually turn to become darken and dim. However, this is what the Quran has explicitly announced through the following verses:

-When the stars become dim.

Chapter77: verse8

-And when the stars darken.

Chapter81: verse2

Lost Stars

When scientists were calculating stars' orbits, they thought that there must be some stars that affect other stars' gravity. They tried very much to find them with telescopes but they could not see these lost stars anywhere. After discovering of black holes, scientists understood why they were calculating lost stars in space.

As we stated before, a black hole is a region of space (not a solid body) into which matter has *fallen* and from which nothing whether material objects or even light itself can escape. If black holes exist, we can never see them directly since they neither emit nor reflect light. They truly live up to their name.

The term "Black Hole" is certainly appropriate for an object such as this. Light –or matter- can *fall* inside it; but nothing whatsoever can escape again[11]. Anyhow, there is a verse in the Quran, which explicitly indicates this falling of the stars:

-No! I swear by the fallings of the stars.

Chapter56: verse75

In this verse, the word "Mawaqi" which is the plural form of the word "Mauqi" means: What is bright or beautiful to our senses may disappear from our ken within a few hours, even though its own existence may continue[12]. Also, according to the Lissan Al-Arab encyclopedia, this word can be translated as "the place of being hidden"[13]. In this verse, there really is a favor for us to dwell upon.

Piercing Star

Scientists believe that the stars, which have big masses, bend space. However, black holes not only bend space, but also pierce it. That is why the scientists call them black holes. It is really impossible to visualize these piercing spaces. But there are black holes in space and they are piercing the space in the dark[14].

The Holy Quran calls black holes "Thaqib". This word means "Penetrating" or "piercing" in Arabic language.[15]. If we ponder carefully, we can perceive the secrets of the universe existing in the Quran:

-And what shall teach thee what is the night-star? The piercing star!

Chapter86: verse2 and 3

Time Dilation in Black Holes

Strong gravitational fields produce a time dilation effect. In the vicinity of a Black Hole, time would pass more and more slowly relative to distant observers the nearer one got to the event horizon. If you could enter a black hole, your son on earth would become much older than you would in one moment. So, in space, time is not passing like our world-time.

An astronaut in this situation would not feel himself living any longer than his natural span. All time processes, clocks etc, would be slowed down by the same amount in his spacecraft. But if he were to return to earth at regular intervals between Black Hole 'trips', he would discover that much greater intervals of earth-time, perhaps thousands of years, would have passed, and he would find a very difficult world to the one he left behind. (But if the present system of money-lending continued, the investment of back-pay at compound interest during these trips would reap enormous dividends!)[16]

If we were exposed to such strange notions about the secrets of the universe 14 centuries ago, we could certainly have not been able to comprehend them, but now, we can easily perceive what the Quran actually intends to say:

- (Whereby) the angels and the spirit ascend unto Him in a Day whereof the span is fifty thousand years!

Chapter70: verse4

It is only recently discovered that there are dim and invisible stars in the universe, which pierce the space; that is, the black holes. As a result, a very important question is escalated here. How could all these wonderful scientific statements be made by an illiterate Muhammed 1400 years ago? Logically, he could have not received any scientific data from a human source. Nothing is remained to say but it must be a revelation from God Almighty. Existence of such scientific inspirations in Holy Quran undoubtedly underscores Muhammed's prophecy. Any right-minded man would feel compelled to bear witness that Holy Quran is the word of Allah and Muhammed (peace be upon him) is His messenger.

Terminology

Accretion disk: A disk shape formed by gas as it spirals into a black hole.

Black hole: A region of space-time formed by the collapse of a massive object, such as a star. A black hole is coined "black" because nothing, even light, can escape its grasp.

Doppler Effect: Relationship between wavelength and speed where shifting of wavelength occurs when the movement of an object is away or towards an observer.

Event horizon: The boundary of a black hole. The point of no return.

General relativity: Albert Einstein's theory that proposes that gravity is a curvature of four-dimensional space-time caused by the presence of matter.

Gravitation: One of the two main processes in a star in which hydrogen is pulled back to the star's center.

Gravitational collapse: When a massive object collapses under its own weight.

Neutron star: The imploded core of a star between 1.4 and 3 times the mass of a sun produced by a supernova explosion.

Nuclear fusion: One of the two main processes in a star in which hydrogen is blown outward from the star's center.

Primordial black hole: A black hole formed in the early universe.

Quantum mechanics: Conservation of energy theory developed from Planck's quantum principle and Heisenberg's uncertainty principle.

Redshift: Reddening of light as it moves away from us, due to the Doppler Effect.

Singularity: The center of a black hole at which matter is crushed to an infinite density, the pull of gravity is infinitely strong, and the space-time curvature becomes infinitely large.

Space-time: Four-dimensional space.

Speed of light: The speed at which light travels (186,282 miles per second). The distance light can travel in one year is called a light year.

White dwarf: A star less than 1.4 the mass of the sun that has exhausted all of its nuclear fuel and has collapsed to a very small size[17].

References

1. Holy Quran, translated by Abdullah Yusuf Ali, Dar Al Arabia, Beirut, 1968.
2. Holy Quran translated by M.H. Shakir, by Ansarian Publication, Qum, the Islamic Republic of Iran, 1993.
3. Holy Quran translated by Arthur J. Arberry, Ansarian Publication, Qum, the Islamic Republic of Iran, 1993.
4. ibid.
5. Holy Quran translated by Marmaduke Pickthall, George Allen and Unwin Ltd., London, Fifth Edition 1969.
6. Black Hole Formation, by John Chang, http://www.rdrop.com/.
7. Pulsars, http://www.astrosun.tn.cornell.edu/.
8. Black Holes in Space, by Patrick Moore and Iain Nicolson, Orbach and Chambers Ltd. 1979.
9. Astronomy in the Physical Universe,

http://mrcohen1.keel.physics.ship.edu/.
10. The Quran and the Last Prophet (in Persian language), by Ayatollah Makarem Shirazi, Dar AL-Kotob Al-Islamiah, Qum, the Islamic Republic of Iran, 1996.
11. Black Holes in Space, by Patrick Moore and Iain Nicolson, Orbach and Chambers Ltd. 1979.
12. Holy Quran, translated by Abdullah Yusuf Ali, Dar Al Arabia, Beirut, 1968.
13. Lissaan Al-Arab, An Arabic-Arabic Dictionary, by Ibn Manzoor, Dar Sader Publishers, Beirut, 1997.
14. Secrets of Universe in Holy Quran, by Sinan Ataseven, http://www.urfa.net/sinan/.
15. A Dictionary of Modern Written Arabic, (Arabic-English) by hans Wehr, McDonald and Evans Ltd., London.
16. Black Holes in Space, by Patrick Moore and Iain Nicolson, Orbach and Chambers Ltd. 1979.
17. Black Hole Formation, by John Chang, http://www.rdrop.com/.

Pulsars and Neutron Stars

The sun is the closest star to our planet earth. It is just an ordinary star in the Milky Way Galaxy, but one in which all life on earth depends. Scientists believe that billions of different kinds of stars exist in the Milky Way Galaxy. We have recently discovered a lot of marvelous celestial bodies, which we did not know anything about them before, such as Quasars, Black Holes and Pulsars.

Pulsars are rapidly rotating neutron stars with periods less than second. As the name implies, a pulsar emits radio pulses at regular intervals. The pulses are detected when the radio beam sweeps through the direction to the earth, much like a lighthouse beacon. Interestingly, the Quran speaks of the pulsars and describes their feature through these verses:

-But Nay! I swear by the stars,
-That run their course and hide themselves[1].

Chapter81: verse15 and 16

How do the pulsars really function? Do they really run their course and hide themselves as the Glorious Quran describes? To know the physical behavior of pulsars, first we have to know about the neutron and correspondingly the binary stars.

Scientific Viewpoint

In 1967, a graduate student in England named Jocelyn Bell was looking at data from a radio telescope and found, much to her surprise, that one radio source was emitting a pulse of radiation every 1.33 seconds. For a while, astronomers thought this might be a signal from intelligent extraterrestrial creatures "little green men". Soon, however, many more of these sources were found, including one in the Crab supernova remnant that emits 30 pulses a second.

Astronomers now believe that these sources, dubbed pulsars, are rapidly spinning neutron stars with strong magnetic fields. The rotating magnetic field produces an electric current, in much the same way that an electric generator operates here on Earth. As the electrons in the current are accelerated, they emit electromagnetic radiation in a sort of conical beam. The radiation can have a broad range of wavelengths, from radio waves through X-rays. Each time that beam sweeps by us, we see a burst of radiation (akin to the flashing of a lighthouse beacon)[2].

What is a Neutron Star?

Neutron stars are very dense and spin very fast and are typically only 10-15 km in radius. Because neutron stars form from burnt-out stars, they do not glow. The collapse of the star causes the matter to be converted into mostly neutrons, and that explains the name neutron star.

Some neutron stars emit radio waves that pulse on and off. These stars are otherwise known as pulsars. Pulsars don't really turn radio waves on and off--it just appears that way to people on Earth because the star is spinning. What happens is that the radio waves only escape from the North and South magnetic poles of the neutron star? If the spin axis is tilted with respect to the magnetic poles, the escaping radio waves sweep around like the light beam from a lighthouse. Far away on Earth, radio astronomers pick up the radio waves only when the beam sweeps across the Earth[3].

How Neutron Stars Form

Black holes and neutron stars form when stars die. While a star is burning, the heat in the star pushes out and balances the force of gravity. When the star's fuel is spent, and it stops burning, there is no heat left to go against the force of gravity. Whatever material is left over collapses in on itself. How much mass the star had when it died determines what it becomes. Stars about the same size as the Sun become white dwarfs, which glow from left over heat. Stars that have about 3 times the mass of the Sun compact into neutron stars. And a star with mass greater than 3 times the Sun's gets crushed into a single point, which we call a black hole[4].

Pulsars

As stated before, pulsars are rapidly rotating neutron stars with periods less than s. The fastest known millisecond pulsar is PSR B1937+21 with a pulsation period of ms. 26 milliseconds. Millisecond pulsars are thought to be sped up by mass dumped onto them by a companion star[5].

Highly energetic electrons spiraling in the magnetic field emit radio beams along the magnetic axis. The pulses are detected when the radio beam sweeps through the direction to the earth, much like a lighthouse beacon.

As their name implies, a pulsar emits radio pulses at regular intervals. The duration of the pulse itself is very short, of order microseconds. The interval between pulses is called the pulse period. Typical pulse periods range from 0.25 to 2 seconds; pulsars with periods in the range of 1 to 10 milliseconds are called millisecond pulsars[6].

Kepler's Laws apply to binary stars also!

Half of all stars in the sky are members of binary systems. The stars orbit in ellipses around a common center of mass. Notice that at any time, the line that connects the two stars must pass through the center of mass.

Each star follows the second law on its own, sweeping out equal areas in equal times within its own orbit.

The pulsar and its companion both follow elliptical orbits around their common center of mass. Each star moves in its orbit according to Kepler's Laws; at all times the two stars are found on opposite sides of a line passing through the center of mass. The period of the orbital motion is 7.75 hours, and the stars are believed to be nearly equal in mass, about 1.4 solar masses. The orbits are quite eccentric. The minimum separation at *periastron* is about 1.1 solar radii; the maximum separation at *apastron* is 4.8 solar radii.

The pulse repetition frequency, or the number of pulses received each second, can be used to infer the radial velocity of the pulsar as it moves through its orbit. When the pulsar is moving towards us and is close to its periastron, the pulses should come closer together; therefore, more will be received per second and the pulse repetition rate will be highest. When it is moving away from us near its apastron, the pulses should be more spread out and fewer should be detected per second.

If the radio beam sweeps past the Earth as the star rotates, we detect a radio pulsar. Some neutron stars may never be seen as pulsars, because their beams do not sweep in the direction of Earth.

Whatever the detailed mechanism, radiation near the neutron star poles produces strong, narrow beams of light which sweep around the sky like a tilted lighthouse. If the earth lies in the path of the beam we see a pulsar. (This idea has the added attraction that it explains why we don't see pulsars in all supernova remnants.)

Quranic Viewpoint

The term "Khunnas" in the previously mentioned verse means "to hide" or "to slink". Also, the term "Kunnas" implies to "to be covered", "to be hidden in one's own place", etc[7].

Having gone over the scientific viewpoint carefully, we can now simply perceive what the Holy Quran intends to state through these verses:

-But Nay! I swear by the stars,
-That run their course and hide themselves.

Chapter81: verse15 and 16

This verse has also been translated as the following:

-No! I swear by the slinkers,
-the runners, the sinkers[8].

This passage directly describes what has occurred in the preceding scientific explanation. Is it not amazing that it could be so exact and precise an explanation long before there was equipment or even the mental capacity for men to understand such things far beyond their vision?

We arrive at this conclusion that this verse describes the status of the particular type of revolving stars, which functionally run their course and hide themselves in their own position. However, this Quranic description of these types of stars exactly complies with what astronomers observed in Steward Observatory in Arizona, 1968:

"These scientists by utilizing the stroboscope observed that this optical star in the crab nebula appears and disappears periodically and twinkles with a pulsation interval of 0.03 s."[9]

Astronomers now believe that all these kinds of sources are the dubbed pulsars, which are rapidly spinning neutron stars with strong magnetic fields.

Are not these verses actually describing the pulsars? Now it is quite easily apprehensible that they are. Once again this author asks the reader, what explanation can account for the deliberate and intelligent assessment of our cosmos from such a remote place in history? No other explanation exists except through divine intervention in the manifestation of the supreme author Allah, as handed down to His prophet Muhammed (peace be upon him and his household) so many 1400 years ago! Allah first commanded our Holy prophet Muhammed "read" and so men have and over the intervening time have repeatedly discovered increasingly complex mysteries of the universe and time unfold in its verses.

What may the existence of such wonderful knowledge in the Quran reveal to us? Undoubtedly, Holy Quran is an unchanged book and miraculous revelation from God Almighty in the sight of those who ponder.

References
1. Holy Quran translated by M.H. Shakir, by Ansarian Publication, Qum, the Islamic Republic of Iran, 1993.
2. Pulsars: Stellar Beacons, http//www.mrcohen1.keel.physics.ship.edu/.
3. http://www.antwrp.gsfc.nasa.gov/.
4. http://www.antwrp.gsfc.nasa.gov/.
5. http:// www.treasure-trove.com/.
6. http://www.astrosun.tn.cornell.edu/.
7. Encyclopedia of Quran (Qaamousse Quran), by Ali Akbar Qurashi, Dar Al-kutub Al-Islamiah. See also: Majma' Al-bayan, An Interpretation of Quran (in Persian), by Abu-Ali Tabarasi, Farahani Publishers, Teheran, 1979.
8. Holy Quran translated by Arthur J. Arberry, Ansarian Publishers, Qum, the Islamic Republic of Iran, 1999.
9. Concepts of Contemporary Astronomy, by Paul Hauge, McGraw-Hill publishers, 1974.

Further Reading
- Measurements of General Relativistic Effects in the Binary Pulsar PSR1913+16, Taylor, J.H., Fowler, L.A. and Weisberg, J.M. 1979, Nature 277, 437.
- The Binary Pulsar: Gravity Waves Exist, Will, C. 1987, Mercury, Nov-Dec, p. 162.
- Gravitational Waves from an Orbiting Pulsar, Weisberg, J.M., Taylor, J.H., Fowler, L.A., 1981, Scientific American Oct, 74.

Sun and Moon and Their Orbits

وَهُوَ ٱلَّذِى خَلَقَ ٱلَّيْلَ وَٱلنَّهَارَ وَٱلشَّمْسَ وَٱلْقَمَرَ كُلٌّ فِى فَلَكٍ يَسْبَحُونَ ﴿٣٣﴾

Dr. Bucaille, the author of the famous book "The Bible, the Quran and Science" states: "When the sun moves through space, there are two options: it can travel just as a stone would travel if one threw it, or it can move of its own accord. The Holy Quran states the latter - that it moves as a result of its own motion:

-And He it is Who created the night and the day, and the sun and the moon. They float each in an orbit[1].
Chapter21: verse33

The Quran uses a form of the word "sabaha" to describe the sun's movement through space. In order to properly provide the reader with a comprehensive understanding of the implications of this Arabic verb, the following example is given.

If a man is in water and the verb "sabaha" is applied in reference to his movement, it can be understood that he is swimming, moving of his own accord and not as a result of a direct force applied to him. Thus, when this verb is used in reference to the sun's movement through space, it in no way implies that the sun is flying uncontrollably through space as a result of being hurled or the like. It simply means that the sun is turning and rotating as it travels.

Now, this is what the Glorious Quran affirms, but was it an easy thing to discover? Can any common man tell that the sun is turning? Only in modern times was the equipment made available to project the image of the sun onto a tabletop so that one could look at it without being blinded. And through this process it was discovered that not only are there spots on the sun but that these spots move once every 25 days. This movement is referred to as the rotation of the sun around its axis and conclusively proves that, as the Quran stated 1400 years ago, the sun does, indeed, turn as it travels through space."

Scientific Viewpoint

The Existence of Sun's Orbit

Ancient people use to believe that the Sun revolves around the earth. Later, Nicholas Copernicus in 1512 laid his Heliocentric Theory of Planetary motion, which placed the sun motionless in the center of the solar system with all the planets revolving around it. Modern science tells us how the sun too is not still, but is in motion.

The sun traveling at roughly 150 miles per second takes about 250 million years to complete one revolution around the center of our Milky Way Galaxy and 25 days to make one complete rotation around its own axis[1].

Orbits of Sun and Moon

Today we know that the Moon revolves around the earth in approximately 29.5 days. The sun also revolves in its own orbit. To understand the sun's orbit, Dr. Bucaille contends that the position of the sun in our galaxy must be considered, and we must therefore call on modern scientific ideas.

Our galaxy, the Milky Way Galaxy, includes one hundred billion stars situated in such a formation that the galaxy is shaped like a disc. This disc turns around its center like a phonograph record or compact disc. Now, it is obvious that when a disc or record turns, any point on the disc would move around and come back to its original position. Similarly, every star in the galaxy moves as the galaxy rotates on its axis. Therefore the stars, which are away from the center of the galaxy, orbit around the axis. The sun is one of those stars.

Dr. Bucaille explains that modern science has worked out the details of the sun's orbit as follows: "To complete one revolution on its own axis, the galaxy and the sun take roughly 250 million years. The sun travels roughly 150 miles per second in the completion of this".

Quranic Viewpoint

See how beautifully and accurately God revealed this phenomenon in the Quran:

-It is He who created the night and the day, and the sun and the moon, all (the celestial bodies) swim along, each in its orbit with its own motion.
Chapter21: verse33

Dr. Bucaille continues: "The above is the orbital movement of the sun that was already referred to in the Quran fourteen centuries ago."

And yet this is a new finding. As Dr. Bucaille reiterates:" The knowledge of the sun's orbit is an acquisition of modern astronomy."

Two verses in the Holy Quran specifically refer to the orbits of the sun and the moon. After mentioning the sun and the moon, God says: "Each one is traveling in an orbit with its own motion" (Quran 21:33; 36:40). How did the author of the Quran know of this? Even after the Quran was revealed, early commentators could not conceive of the orbits of the sun and moon. The tenth century commentator Tabari could not explain this so he said, "It is our duty to keep silent when we do not know".

Dr. Bucaille points out the incredibility of these suppositions outside modern science where when he states: "This shows just how incapable men were of understanding this concept of the sun's and moon's orbit".

From this, it is clear that if the most glorious Holy Quran was expressing an idea already known to the people, the commentators would have easily understood it. But this, as Dr. Bucaille explains was "a new concept that was not to be explained until centuries later".

This confirms what God said to his prophet, on whom be peace:

-This is of the tidings of the Unseen, which we inspire in you (Muhammed). Neither you nor your people knew it before this.
Chapter11: verse49

The Sun and Moon Move With Their Own Motion

The Quran makes the following statement about the sun and the moon: "Each one is traveling in an orbit with its own motion" (Quran 21:33; 36:40).

Why did the Holy Quran say that the sun and moon move with their own motion? And, if that is true, where did the author of the Quran get this information?

The fact is that the sun and moon rotate on their axes and are in part animated by this rotating motion. The phrase "traveling with its own motion" in the verses quoted above is a translation of the Arabic verb 'yasbahoon'. This could also be translated 'they swim.' In that case, the verse would read that the sun and the moon, "each swim in its own orbit." Those who translate the verse this way explain that the term swim refers to movement with one's own internally generated force. Furthermore, the movement of a swimmer is graceful, measured, and smooth. This is a very fitting description for the movement of the stars and planets including the sun and the moon.

After describing the scientific data concerning the rotation of the sun and the moon, Dr. Bucaille says: "These motions of the two celestial bodies are confirmed by the data of modern science, and it is inconceivable that a man living in the seventh century A.D.... could have imagined them".

It is also amazing that the Quran uses a different term for the movement of the clouds and the mountains (see Quran 27:88). Obviously, the clouds and mountains are driven by external forces. The cloud is driven by the wind and the mountains move with the rotation of the earth. The sun and moon, however, move with their own motion, and therefore the Quran uses a peculiar term "they swim" to refer to their smooth, graceful, self-propelled movement.

How did the author of Holy Quran know enough to make this choice of words that will reflect a modern scientific truth? Because the essence of the revelations of Holy Quran are not from a man, who is limited in his perceptions, observations, and deduction...The Quran is no less than a revelation from God[3].

References
1. Holy Quran translated by Marmaduke Pickthall, George Allen and Unwin Ltd., London, Fifth Edition 1969.
2. http://www.gsfc.nasa.gov/.
3. The Bible, The Quran and Science (Le Bible, le Coran et la Science)," The Holy Scriptures Examined in the Light of Modern Knowledge, by Dr. Maurice Bucaille, French Physician, Seghers, Paris, 1987, English version published by North American Trust Publication, 1978, page: 161,162 and 163.

Inherent Diversity Between Sun and Moon

وَجَعَلْنَا سِرَاجًا وَهَّاجًا ۝

تَبَارَكَ الَّذِى جَعَلَ فِى السَّمَاءِ بُرُوجًا وَجَعَلَ فِيهَا سِرَاجًا وَقَمَرًا مُّنِيرًا ۝

هُوَ الَّذِى جَعَلَ الشَّمْسَ ضِيَاءً وَالْقَمَرَ نُورًا وَقَدَّرَهُ مَنَازِلَ لِتَعْلَمُوا عَدَدَ السِّنِينَ وَالْحِسَابَ مَا خَلَقَ اللَّهُ ذَٰلِكَ إِلَّا بِالْحَقِّ يُفَصِّلُ الْآيَاتِ لِقَوْمٍ يَعْلَمُونَ ۝

The sun produces energy and light, which is vital for the human to survive on earth. The moon reflects a small part of the sunlight from its surface, which is visible at nights.

The matter of fact is that the sun and moon are inherently different from each other. Although, this has been discovered after the invention of telescope and by the means of modern equipment. In the course of history, man believed in a lot of fanciful and whimsical myths regarding the sun and the moon, whereas, the Holy Quran through some very interesting verses implicitly clarified the inherent different functions of the sun and the moon 1400 years, as follows:

-*And We appointed a blazing lamp*[1].

Chapter78: verse13

-Blessed be He who has set in heaven constellations, and has set among them a lamp, and an illuminating moon[2].

Chapter25: verse61

-It is He Who made the sun a radiance, and the moon a light...[3]

Chapter10: verse5

To comprehend the purport of these glorious verses, we have to review what scientists present in this regard. Then, we will revisit the Quranic viewpoint.

Scientific Viewpoint

The Sun

Our sun is just an ordinary star like many others in the night sky, but it is very important to us. Without it the earth would be dark and cold and nothing could live here. It is integral to our total and absolute existence.

The sun and the moon are different from each other not only in terms of size, but also in terms of function. The sun generates light, but the moon does not. The moon merely reflects the light coming from the sun. This is a fact learned early in modern educational systems.

The sun is a huge, spinning mass of very hot gas, mostly hydrogen, but it does not burn like an ordinary fire. At the center of the sun, the temperature is so high and the pressure so strong that atoms of hydrogen gas combine to make helium gas. This is the same process that occurs in a hydrogen bomb and it produces a great deal of energy. In doing this, the sun loses 4000 million kg (4million tons) of material every second. However, the sun is so huge that even at this rate, it has taken 4500 million years to use up half its hydrogen. It still has enough left, however to keep shining steadily for another 5000 million years[4].

The sun sends out its energy as heat and light in all directions. On earth every square meter directly facing the sun, receives more than one kilowatt of energy (The amount given out by a single bar electric fire).

We are fortune that the sun is exactly the way it is. If it were different in almost anyway, life would certainly never have developed on earth[5]. It is the unique combination of a multitude of factors that makes our sun capable of supporting the vast area of life it does on our planet.

The Moon

The earth's only natural satellite- the moon, the one planetary body other than the earth, which we have visited. This astronomical body easily observed from the earth, the moon, is second only to the sun in size and brilliance from our vantage point. In reality, the moon is relatively small; 3476 km in diameter, it is only about 27% of the size of the earth. It shines only by reflected sunlight and is a poor reflector. On average, only about 7.3% of the visible light that falls upon the surface bounces back into space[6].

Quranic Viewpoint

The word "Moneer" in the following verse means something bright and illuminating, which gives light but does not produce its light itself:

-Blessed be He who has set in heaven constellations, and has set among them a lamp, and an illuminating moon.
Chapter25: verse61

Could A man or woman in the seventh century, have known about this fine distinction between the sun and the moon? To such a person, the two would appear as a greater light and a lesser light. Such a person would observe that the greater light lights up the day and the lesser light lights up the night. And this indeed is how the sun and the moon were described in previous books.

Dr. Bucaille, the famous author of the book" The Bible, the Quran and Science" comments in his book:

"...The Bible, describing the creation, says: "God made two great lights the greater light to govern the day and the lesser light to govern the night" (Genesis 1: 16). The author of the Quran however, was aware that this comparison between the sun and the moon is not adequate. Therefore the Quran does not refer to them as being a greater and a lesser light. The Quran says:

-God is the One who made the sun a shine and the moon a light.
Chapter10: verse5

Commenting on this, Dr. Bucaille says: "Whereas the Bible calls the sun and moon 'lights', and merely adds to one the adjective 'greater' and to the other 'lesser', the Quran ascribes differences other than that of dimension to each respectively"[7].

Similarly, the Quran says:

-Blessed is the One Who placed the constellations in heaven and placed therein a lamp and a moon giving light.
Chapter25: verse61

Once again, the difference between the sun and the moon is noted. The sun is called a lamp, and the moon is called an object giving light. Again in the Quran God says:

-He "made the moon a light" and "made the sun a lamp".
Chapter71: verse15-16

Furthermore, God calls the sun a "blazing lamp":

-And We appointed a blazing lamp.
Chapter78: verse13

This term, which is used for the sun is never used for the moon in Holy Quran. In all of these verses, God expresses the notion that the sun and the moon are "not absolutely identical lights"[8].

Dr. Bucaille draws his conclusions from what he found in the Quran about the sun and the moon: "What is interesting to note here is the sober quality of the comparisons, and the absence in the text of the Quran of any elements of comparison that might have prevailed at the time and which in our day would appear as phantasmagorical".

In essence, "There is nothing in the text of Holy Quran that contradicts what we know today about these two celestial bodies"[9].

Fourteen centuries ago, no one could discriminate the sun's light and the moonlight and apprehend their functional discrepancy, but the Holy Quran made it clear that the sun is a lamp, which produces light and the moon gives (reflects) light.

References
1. Holy Quran translated by Arthur J. Arberry, Ansarian Publication, Qum, the Islamic Republic of Iran, 1999.
2. ibid.
3. ibid.
4. Exploration of the Universe, by Abell, Morrison, Wolff, Saunders College publishing, USA, 1987.
5. From "Information Summaries", National Aerospace and Space Administration (NASA), June 1991.
6. Philips' Moon Map, Philips Publication, London.

7. The Bible, The Quran and Science (Le Bible, le Coran et la Science)," The Holy Scriptures Examined in the Light of Modern Knowledge, by Dr. Maurice Bucaille, French Physician, Seghers, Paris, 1987, English version published by North American Trust Publication, 1978.
8. ibid.
9. ibid.

Revolution of Sun

وَٱلشَّمْسُ تَجْرِى لِمُسْتَقَرٍّ لَّهَا ذَٰلِكَ تَقْدِيرُ ٱلْعَزِيزِ ٱلْعَلِيمِ ۝

ٱللَّهُ ٱلَّذِى رَفَعَ ٱلسَّمَٰوَٰتِ بِغَيْرِ عَمَدٍ تَرَوْنَهَا ثُمَّ ٱسْتَوَىٰ عَلَى ٱلْعَرْشِ وَسَخَّرَ ٱلشَّمْسَ وَٱلْقَمَرَ كُلٌّ يَجْرِى لِأَجَلٍ مُّسَمًّى يُدَبِّرُ ٱلْأَمْرَ يُفَصِّلُ ٱلْآيَٰتِ لَعَلَّكُم بِلِقَآءِ رَبِّكُمْ تُوقِنُونَ ۝

It might be expected that the sun, a typical star, which is in motion; just as the other stars are. As we know, the sun is a member of the Milky Way Galaxy, a system of a hundred thousand stars, which revolve around the Milky Way Galaxy.

According to the first following verse, the most Holy Quran gives an end to the sun for its revolution and a destination place. It also provides the moon with a place of rest.

-And the sun runneth on unto a resting-place for him. That is the measuring of the mighty, the wise.

Chapter36: verse38

-And (Allah) compelled the sun and the moon to be of service, each runneth unto an appointed term[1].

Chapter13: verse2

To understand the possible meanings of these statements, we must remember what modern knowledge has to say about the revolution of the stars in general and the Sun in particular, and (by extension) the celestial bodies that automatically followed its movement through space, among them the Moon.

But interestingly enough, we find this knowledge in a book, which was recited by Muhammed (Peace be Upon Him) the prophet of Islam 1400 years ago. How could this astonishing knowledge have been presented by an illiterate man himself, whose job was watching the sheep and escorting caravans of goods in the dessert of Arabia? Absolutely, it could not have been his words but a revelation from Allah, the creator of everything in existence.

Scientific Viewpoint

The most luminous stars in the galaxy, including our sun lie in a disk, which is flat, like a pancake and is rotating. Planets also follow the sun through the space. The Sun is a star that is roughly 4.5 billion years old, according to experts in astrophysics. It is possible to distinguish a stage in its evolution, as one can for all the stars.

At present, the Sun is at an early stage, characterized by the transformation of hydrogen atoms into helium atoms. Theoretically, this present stage should last another 5.5 billion years according to calculations that allow a total of 10 billion years for the duration of the primary stage in a star of this kind. It has already been shown, in the case of these other stars, which this stage gives way to a second period characterized by the completion of the transformation of hydrogen into helium, with the resulting expansion of its external layers and the cooling of the Sun.

In the final stage, its light is greatly diminished and density considerably increased; this is to be observed in the type of star known as a 'white dwarf' (See "Death of the Universe").

The above dates are only of interest in as far as they give a rough estimate of the time factor involved. What is worth remembering and is really the main point of this discussion, is the notion of an evolution. Modern data allow us to predict that, in a few billion years, the conditions prevailing in the solar system will not be the same as they are today. Like other stars whose transformations have been recorded until they reached their final stage, it is possible to predict an end to the Sun.

The sun partaking of this general rotation of the galaxy is moving toward the constellation Hercules, at a speed of almost 72000 km an hour or 250 km/s to complete its orbit about the galactic center [2]. The sun's gravitational pull on the planets is so strong that they follow right along. The solar system is the part of the Milky Way Galaxy, which is rotating too. It makes a full rotation about every 250 million years.

Over 400 years ago, Copernicus advanced a heliocentric -Sun Centered- theory. He suggested that the retrograde motion of the planets could be readily explained if the sun, rather than the earth, was at the center of the universe; that the earth is a planet; and that the planets move around the sun in circles.

Copernicus theory, although it put the sun instead of the earth at the center of the solar system (and, for then, the universe) still assumed that the orbits of celestial bodies had to follow such perfect orbits, shows that Copernicus had not been able to break away entirely from the old ideas. The so-called Copernican revolution, in the sense that our modern approach to science involves a comparison of nature and understanding really occurred later[3].

Quranic Viewpoint

-And the sun runneth on unto a resting-place for him. That is the measuring of the mighty, the wise.

Chapter 36: verse 37

The verse quoted above, referred to the Sun running its course towards a place of its own.

Fourteen centuries ago, in the time of revelation of this verse, while everyone of that time thought that the earth is at the center of the universe, Holy Quran through this glorious verse, spoke about the motion of the sun and its resting place.

As we stated previously, modern astronomy has been able to locate it exactly and has even given it a name, the Solar Apex: the solar system is indeed evolving in space towards a point situated in the Constellation of Hercules (alpha lyrae) whose exact location is firmly established; it is moving at a speed already ascertained at something in the region of 12 miles per second.

This extensive astronomical data deserves to be mentioned in relation to this verse from the Quran, since it is supported by modern scientific data[4].

Existence of such astonishing knowledge contained in the historic Holy Quran is in accordance with the modern provable scientific discoveries. This actually strikes a responsive chord among the open-minded readers with the divinity of this book, especially while noticing the era and the place in which the Quran appeared.

Terminology
Hercules (HER): Strongman - One of the original 48 constellations listed by Ptolemy in the second century AD and a large constellation of the northern hemisphere, lying between Lyra and Corona Borealis. Hercules is best seen during the summer. It is represented by the figure of the Greek hero Hercules in a kneeling position. The stars of the constellation are of third magnitude or dimmer. Hercules contains a globular cluster, called Messier 13, of more than 50,000 stars. This cluster, about 34,000 light-years from the earth, can be seen by the naked eye [5].

References
1. Holy Quran translated by Marmaduke Pickthall, George Allen and Unwin Ltd., London, Fifth Edition 1969.
2. StarFinder Series (c) State of Maryland Department of Education, distributed by Film Ideas.
3. Exploration of the Universe, by George Abell, Hot, Rinehart and Winston Inc., USA, 1969.
4. The Bible, The Quran and Science (Le Bible, le Coran et la Science)," The Holy Scriptures Examined in the Light of Modern Knowledge, by Dr. Maurice Bucaille, French Physician, Seghers, Paris, 1987, English version published by North American Trust Publication, 1978.
5. Encarta 2000 Encyclopedia.

Composition of Meteors

يُرْسَلُ عَلَيْكُمَا شُوَاظٌ مِن نَارٍ وَنُحَاسٌ فَلَا تَنتَصِرَانِ ۝

Have you ever seen a shooting meteor trailing in the night sky? Meteor showers are one of the most spectacular phenomena, which are observable in the sky with the naked eyes.

A very large number of meteoroids enter the Earth's atmosphere each day adding up to more than a hundred tons of material. But they are almost all very small, just a few milligrams each. Only the largest ones ever reach the surface to become meteorites. Therefore, they could scarcely jeopardize the life on earth, thanks to our protective atmosphere.

So, you do not need to be concerned at all. These heavenly stones do not survive in most cases to impact on the ground, since they burn up in the sky while colliding with the atmosphere. (See "The Protective properties of Atmosphere").

What are these flaming molten stones made of? Scientists believe that the meteorites are mostly composed of rock and metal like iron and copper[1]. Interestingly enough, this is what the Quran has explicitly stated concerning the composition of meteors and meteorites 1400 years ago:

-*Against you shall be loosed a flame of fire [of iron], and molten brass [copper]; and you shall not be helped*[2].

Chapter55: verse35

The reference should be made to the lately discovered scientific facts in this regard.

Scientific Viewpoint

Meteorites are dusts and rocks in space. They come from two main sources, comets and asteroids.

Comets lose dust and fragments if they come close to the sun and evaporate; asteroids lose fragments if they collide together. Around 220000 tons of such material enters the earth's atmosphere each year. Once in the atmosphere, the friction between the meteoroid and air molecules often produces a brief trail of light called a meteor or shooting star. If the meteoroid does not burn up completely but survives to hit the ground, it is called a meteorite. The largest found meteorite (Hoba, in Namibia) weighs 60 tons.

The average meteoroid enters the atmosphere at between 10 and 70 km/sec. But all but the very largest are quickly decelerated to a few hundred km/hour by atmospheric friction and hit the Earth's surface with very little fanfare. However meteoroids larger than a few hundred tons are slowed very little; only these large (and fortunately rare) ones make craters.

Meteorite Types

There are three main types of meteorite:

1- Stony (made mainly of rock)
2- Metallic (made mainly of iron)
3- Stony-Metallic (made of both rock and iron)

A meteor, more commonly termed a shooting star, is the streak of light produced when a meteoroid burns up in the earth's atmosphere. On a clear moonless night about 10 an hour can normally be seen. The number of visible meteors peaks at around 4 a.m. Since the observer is on the side of the earth that is heading into the dust. Meteors are best seen when they appear in a meteor shower, which occurs when the earth passes through a trail of dust left by a comet.

A comet is a relatively small object that follows a long, eccentric orbit around the sun and consists of dust and ice. Most comets orbit far beyond the planets at distances of up to 100000 astronomical units (about 15000 billions km). If a comet's orbit brings it close to the sun, the comet partially vaporizes, producing a bright head of dust and gas and one or more trails.

An asteroid is a small rocky object that orbits the sun and is smaller in size than a planet. Asteroids may be stony carbonaceous or metallic in composition and they range in size from about 1000 km (620 miles) across, down to the smallest dust particles. Asteroids are found mainly in the asteroid belt, which lies between the orbits of the planets Mars and Jupiter.

Of all the meteorites examined, 92.8 percent are composed of silicate (stone), 5.7 percent are composed of iron and nickel, the rest are a mixture of the three materials. Stony meteorites are the hardest to identify since they look very much like terrestrial rocks.

Since asteroids are material from the very early solar system, scientists are interested in their composition.

A good example of what happens when a small asteroid hits the Earth is Barringer Crater (a.k.a. Meteor Crater) near Winslow, Arizona. It was formed about 50,000 years ago by an iron meteor about 30-50 meters in diameter. The crater is 1200 meters in diameter and 200 meters deep. About 120 impact craters have been identified on the Earth, so far.

Quranic Viewpoint

The Quran through the verse 33 and 35 of chapter 55 states that if Human beings and jinn (demons) could penetrate the regions of heavens, (see "Conquest of Space by Human") they would encounter the flame of fire and molten metals like iron and copper:

-Against you shall be loosed a flame of fire [of iron], and molten brass [copper]; and you shall not be helped.

In this verse, there are two words, which are deemed to be the Quranic evidence for the discussed scientific viewpoint. The word "Shuwaaz" is the plural form of the word "shaizah", which means "molten iron"[3]. Also, according to the Arabic encyclopedias and Quranic interpretations, the word "Nou'haass" means "copper"[4].

Needless to say, this Quranic statement is in compliance with the respective scientific discoveries concerning the composition of meteors and meteorites, which were discussed already.

Undoubtedly, neither Muhammed (peace be upon him) nor any one else could have been aware of the composition of heavenly bodies even many centuries later. Absolutely, the existence of such a precise knowledge in the Quran proves its authenticity as a divine revelation from the standpoint of science today.

References

1. Travel to Space, a series of "Quran and Nature" (in Persian), by Ab-

dulkareem Biazaar Shirazi, Be'sat Publishers institute, Teheran, 1971.
2. Holy Quran translated by Arthur J. Arberry, Ansarian Publishers, Qum, the Islamic Republic of Iran, 1999.
3. Travel to Space, a series of "Quran and Nature" (in Persian), by Abdulkareem Biazaar Shirazi, Be'sat Publishers institute, Teheran, 1971.
4. Aghrab Al-mawarid, An Arabic-Arabic Dictionary by Allamah Sa'eed Sa'eed Al-Khouli Al-Shartoon Al-Lubnaani.

Further Reading
- Meteors, Meteorites and Impacts, http://www.hawastsoc.org/.
- Meteoroids and Meteorites, by Rosanna L. Hamilton, http://www.solarviews.com/, 1999.
- Beatty, J. K. and A. Chaikin. The New Solar System. Massachusetts: Sky Publishing, 3rd Edition, 1990.
- Maran, Stephen P., The Astronomy and Astrophysics Encyclopedia. New York: Van Nostrand Reinhold, pp. 430-445, 1992.
- Seeds, Michael A. Horizons. Belmont, California: Wadsworth, 1995.

Motion of Earth

وَتَرَى ٱلْجِبَالَ تَحْسَبُهَا جَامِدَةً وَهِىَ تَمُرُّ مَرَّ ٱلسَّحَابِ صُنْعَ ٱللَّهِ ٱلَّذِىٓ أَتْقَنَ كُلَّ شَىْءٍ إِنَّهُۥ خَبِيرٌۢ بِمَا تَفْعَلُونَ ﴾

Today, one of the most widely known scientific facts even to the average man is the motion of earth around the sun. This knowledge has come to us only in recent centuries after the invention of telescope by which human became able to observe the motion of planets around the sun.

As mentioned in the former articles, the ancient Greeks began to explain the motions of the planets by making theoretical models of the geometry of the solar system. One of the earliest and greatest philosophers, Aristotle, lived in Greece about 350 BC He summarized the astronomical knowledge of his day into a qualitative cosmology that remained dominant for 2800 years. On the basis of what seemed to be very good evidence_ what he saw_ Aristotle thought and actually believed that he knew that the earth was at the center of the universe and that the planets, the sun and the stars revolved around it.

According to Aristotle the universe was made up of a set of 55 celestial spheres that fit around each other; each has rotation as its natural motion. Each of the heavenly bodies was carried around the heavens by one of the spheres. The motion of the spheres affected each other and combined to account for various observed motions of the planets, including revolution around the earth, retrograde motion and motion above and below the ecliptic. The outermost sphere was that of the fixed stars, beyond which lay prime mover, premium mobile that caused the general rotation of the stars overhead.

Aristotle's theories dominated scientific theories for almost two thousand years, until the renaissance. Unfortunately, most of his theories were far from what we now consider to be correct, so we tend to think that the widespread acceptance of Aristotelian physics impeded the development of science[1].

Despite these prevalent theories in the seventh century, the Holy Quran presented a verse through which the motion of earth is implicitly stated in a very deliberate way:

-Thou seeth the mountains and thinkest them firmly fixed, but they shall pass away as the clouds pass away; (such is) the artistry of Allah who disposes of all things in perfect order, for He is well-acquainted with all that ye do[2].

Chapter27: verse88

However, one may question that there is no clear sign of the earth's motion being mentioned in this verse. Let us see how this verse is deserved to be in relation with such an issue.

Scientific Viewpoint

Earth revolves around the sun in 365 days, 6 hours, 9 minutes with reference to the stars. Its mean orbital speed is about 100,000 km per hour. The 6 hours, 9 minutes adds up to about an extra day every fourth year, which is designated a leap year, with the extra day added as February 29th. Earth's orbit is elliptical and reaches its closest approach to the sun (perihelion) on about January fourth of each year.

Quranic Viewpoint

The preceding verse mentions that the mountains are in motion, even though they seem to be fixed. Absolutely, the motion of mountains would be meaningless, without the motion of the other lands to which they are attached. Obviously, the clouds and mountains are driven by external forces. The cloud is driven by the wind and the mountains move with the rotation of the earth. Therefore, the verse would intend to mean as follows: "The lands altogether are rapidly in motion, like the motion of clouds". The interesting point in this verse is likening the motion of the earth to the motion of clouds.

Two points can be deduced from this wonderful simile:

Firstly, from the standpoint of the world's literature, the rapid motion has generally been likened to the motion of clouds. Secondly, the motion of clouds is extremely tranquil and intangible. Accordingly, the motion of earth is also quite tranquil and without any sensible disturbance[3]. Consequently, we can arrive at this conclusion that the context of this statement refers to the motion of earth.

Comparing the dominant theories about the situation of earth in the universe in the seventh century and what the Quran implies apropos this phenomenon makes us admit the authenticity of Holy Quran as a revelation from God Almighty.

References

1. The Bible, The Quran and Science (Le Bible, le Coran et la Science)," The Holy Scriptures Examined in the Light of Modern Knowledge, by Dr. Maurice Bucaille, French Physician, Seghers, Paris, 1987, English version published by North American Trust Publication, 1978.
2. Holy Quran translated by Marmaduke Pickthall, George Allen and Unwin Ltd., London, Fifth Edition, 1969.
3. The Quran and the Last Prophet (in Persian), by ayatollah Makarem Shirazi, Dar AL-Kotob Al-Islamiah, Qum, the Islamic Republic of Iran, 1996.

Trundling Earth

وَٱلْأَرْضَ بَعْدَ ذَٰلِكَ دَحَىٰهَا ﴿٣٠﴾

We have affectionately termed the world we live in "the earth". We know this planet best. A planet is simply a huge ball of rock or even of liquid and gases, which circles a star. Our planet is made of rock and the star it circles around, is the sun, which has nine planets. The earth is the third planet from the sun. Like the other planets, it moves around the sun in a huge circle called orbit, spinning all the time.

As already mentioned, one of the earliest and greatest Greek philosophers, Aristotle, lived in 350 BC He summarized the astronomical knowledge of his day into a qualitative cosmology that remained dominant for 2800 years. Based on what seemed to be very good evidence, his observations, Aristotle thought and actually speculated that the earth was at the center of the universe and that the planets, the sun and the stars revolved around it.

Aristotle proposed that the universe was comprised of a set of 55 celestial spheres that fit around each other; each has rotation as its natural motion. Each of the heavenly bodies was transported around the heavens by one of the spheres. The motion of the spheres affected each other and combined to account for various observed motions of the planets, including revolution around the earth, retrograde motion and motion above and below the ecliptic. In his opinion, the outermost sphere was that of the fixed stars, beyond which lay prime mover, premium mobile that caused the general rotation of the stars overhead.

Aristotle's theories dominated scientific theories for almost two thousand years, until the Renaissance. Unfortunately, most of his theories were far from what we now consider correct, so we tend to think that the widespread acceptance of Aristotelian physics impeded the development of science[1].

However, 1400 years back, when the Aristotle's theory was the dominant theory in the world, the Quran clearly described the earth spinning through the verse30 of chapter79:

-And after that He expanded the earth by trundling[2].

Chapter79: verse30

Although, the fact that the earth is spinning all the time is now widely accepted fact, except for Quranic references, no one truly understood this concept until the discoveries that came hundreds of years after the Quran was compiled, during the Renaissance. To understand the significance of the above Quranic verse, the reference must be made to the current pertinent scientific viewpoint.

Scientific Viewpoint

Our calendar is based on the earth movements. A day is the time the earth takes to spin around once and it takes a year (365 1/4 days) to circle the sun.

The earth rotates on its axis relative to the sun every 24.0 hours mean solar time, with an inclination of 23.5 degrees from the plane of its orbit around the sun. Mean solar time represents an average of the variations caused by the earth's non-circular orbit. Its rotation relative to the other stars (sidereal time) is 3 min. and 56.55 sec. shorter than the mean solar day, the equivalent of one solar day per year. Forces associated with the rotation of earth cause it to be slightly oblate, displaying a bulge (oblateness) at the equator[3].

Precession of the Earth Axis

The moon's gravity, primarily, and to a lesser degree the sun's gravity acting on Earth's oblateness tries to move earth's axis perpendicular to the plane of earth's orbit. However, due to gyroscopic action, Earth's poles do not "right themselves" to a position perpendicular to the orbital plane. Instead, they precess at 90 degrees to the force applied. This precession causes the axis of Earth to describe a circle having a 23.5 degree radius relative to a fixed point in space over about 26,000 years, a slow wobble reminiscent of the axis of a spinning top swinging around.

As the earth spins, the iron core acts like a giant magnetic field. This magnetism is used to find the headings on a compass. The earth's magnetic field also traps tiny particles from the sun in belts of radiation round the planet.

Quranic Viewpoint

Hundreds of years before the discovery of earth's rotation around itself, Holy Quran had clarified it in the verse 30 of chapter 79:

-And after that He (Allah) expanded the earth by trundling.

Interestingly enough, in sermon 90 of the book "Nahjul Balagha (The Vivid way)", we encounter this wonderful statement made by Imam Ali (Peace Be Upon Him) 1400 years ago concerning the genesis of earth:

-And the earth became tranquil while it was trundling in the waving ocean!

The evidence of the verse is the word "dahaha". This word is derived from the verb "dahawa", which means "to expand something by trundling", "to spread something by rolling" etc. Arabic is a language, which is able to represent some various ideas simultaneously by one word, including the respective pronouns, tense and objects. The word "dahaha" consists of two words: "daha" and "ha". "Daha" is a simple past tense transitive verb and "ha" is the object pronoun, which grammatically refer to the earth. Consequently, the word "dahaha" could literally be translated as "(He) expanded it (the earth) by trundling".[4]

Concurrently, this word also implies to the revolutionary motion of the earth, since every trundling thing is actually transferred from one place to another one.

However, this wonderful statement in Holy Quran about the earth's trundling motion leads us to this fact that Holy Quran must have had a non-human source i.e. Allah, the Almighty Himself is the author of Holy Quran. Indeed, an infinite wisdom exists behind the Quran's simple but pithy words. The Holy Quran is the perpetual miracle, which presents the best guidelines to humanity in the sight of those who reflect.

References

1. The Bible, The Quran and Science (Le Bible, le Coran et la Science)," The Holy Scriptures Examined in the Light of Modern Knowledge, by Dr. Maurice Bucaille, French Physician, Seghers, Paris, 1987, English version published by North American Trust Publication, 1978.
2. Holy Quran translated by Muhammed Mahdi Fulaadvand (in Persian), Dar Al-Quran Al-Kareem, Teheran, 1995.
3. Exploration of the Universe, by George Abell, Hot, Rinehart and Winston Inc., USA, 1969.
4. Ahsan Al-Hadith, An Encylopedia of Holy Quran, by Ali Akbar Qurashi/See also: Al-Meezaan, An Interpretation of Quran, (in Persian) Vol. 40, by Muhammed Hussein Tabataba'i, Muhammadi Publishers, Teheran, 1982; and Majma' Al-bayan, An Interpretation of Holy Quran (in Persian), by Abu-Ali Tabarasi, Farahani Publishers, Teheran, 1979.

Spherical Earth

فَلَا أُقْسِمُ بِرَبِّ الْمَشَارِقِ وَالْمَغَارِبِ إِنَّا لَقَادِرُونَ ۝

يَا مَعْشَرَ الْجِنِّ وَالْإِنسِ إِنِ اسْتَطَعْتُمْ أَن تَنفُذُوا مِنْ أَقْطَارِ السَّمَاوَاتِ وَالْأَرْضِ فَانفُذُوا ۚ لَا تَنفُذُونَ إِلَّا بِسُلْطَانٍ ۝

خَلَقَ السَّمَاوَاتِ وَالْأَرْضَ بِالْحَقِّ ۖ يُكَوِّرُ اللَّيْلَ عَلَى النَّهَارِ وَيُكَوِّرُ النَّهَارَ عَلَى اللَّيْلِ ۖ وَسَخَّرَ الشَّمْسَ وَالْقَمَرَ ۖ كُلٌّ يَجْرِي لِأَجَلٍ مُّسَمًّى ۗ أَلَا هُوَ الْعَزِيزُ الْغَفَّارُ ۝

أَلَمْ تَرَ أَنَّ اللَّهَ يُولِجُ اللَّيْلَ فِي النَّهَارِ وَيُولِجُ النَّهَارَ فِي اللَّيْلِ وَسَخَّرَ الشَّمْسَ وَالْقَمَرَ كُلٌّ يَجْرِي إِلَىٰ أَجَلٍ مُّسَمًّى وَأَنَّ اللَّهَ بِمَا تَعْمَلُونَ خَبِيرٌ ۝

Today we can look at a globe and know that the earth is somewhat like a ball or a sphere. The Holy Quran makes certain statements that led Muslim scientists to understand, long before their European counterparts, that the earth is spherical. When Europe was in the dark ages thinking that the earth was flat, Muslim students were using globes for studying the earth in Islamic universities[1].

The fact that the earth is spherical is now so commonplace that there is seemingly no need for any scientific proof. Indeed, modern man has in fact had the most positive proof available: seeing ourselves from the vantage point of outer space.

Quranic Viewpoint

Since it was not the purpose of the Holy Quran to teach science, the Quran did not need to state explicitly that the earth is spherical in shape (or more precisely, a geoid). But some of what the Quran says, stimulates you to think of the world as a globe. Take, for example, the following verses:

-But nay! I swear by the Lord of the Easts and the Wests that We are certainly able...[2]

Chapter70:verse40

This verse says "The Easts and the Wests of earth". This statement would be correct only if the earth were spherical. If it were flat, there would always be only one designated east and one area designated west. Since the earth is spherical, any point of its surface is considered as a west comparing to the eastern locations and is assumed as an east comparing to the western points.

Therefore, the number of the easts and the wests is equal to the number of the whole points on earth. Consequently, the words "easts and wests" implicitly, prove that the earth is spheroid[3].

In the time of revelation of Quran and also many centuries later, no one was aware that the earth is spherical. So people thought that there are only one east and one west. But Quran strongly stated that many easts and wests occur on earth.

Another verse that should command our attention is:

-O assembly of Jinns and Men, if you can penetrate regions of the heavens and the earth, then penetrate them! You will not penetrate them save with a Power.

Chapter55: verse33

The word "Aqtar" in Arabic means "diameters" which refers to the length of a straight line that goes from one side of a round object to another, through the center of the object[4]. This simple word easily leads us to this idea that we have to pass through the diameters of a spherical object to penetrate into space and the earth itself as well. Therefore, we can come to this conclusion that the fact of spherical earth is implicitly represented in this verse around 1400 years ago.

These verses are also considerable in this regard:

-Have you not seen how God merges the night into the day and merges the day into the night?

Chapter 31: verse 29

Another verse tells us that God coils the day and night around:

-He coils the night upon the day and He coils the day upon the night.

Chapter 39: verse 5

The word 'coils' in the verse above is a translation of the Arabic verb "kawwara", which is used in describing the action of coiling a turban around the head. For a better understanding of this statement, readers of the Quran had to think of the earth as a sphere.

To fully appreciate the above two statements in the Quran, try this experiment at home: You need a flashlight and a globe. Take these items into a dark room. Using the flashlight to simulate the light of the sun, shine the light upon the globe. Notice that only one half of the globe is lighted up. The other half is in darkness. Half the world is having day, the other half is having night. Now, recall that the earth is continuously rotating on its axis and will go around completely in twenty-four hours. Slowly turn the globe around to simulate this rotation. Notice that as the globe turns, the day is going around the globe to light up the other half of the world. The night is also going around the globe to give rest to the other half of the world.

The day and night are perpetually coiling around the earth with some degree of interpenetrating. This is exactly how it appears to astronauts during their space flights.

Dr. Bucaille makes the following remark: "This process of perpetual coiling, including the interpenetrating of one sector by another, is expressed in the Quran just as if the concept of the earth's roundness had already been conceived at the time, which was obviously not the case[5]."

How do we explain the presence of this knowledge in the Quran? This obviously did not reflect the level of common knowledge of the time, but was inspirationally helpful in stimulating Muslim scientists to conceive of the earth in its real shape.

References
1. The Bible, The Quran and Science (Le Bible, le Coran et la Science)," The Holy Scriptures Examined in the Light of Modern Knowledge, by

Dr. Maurice Bucaille, French Physician, Seghers, Paris, 1987, English version published by North American Trust Publication, 1978.
2. Holy Quran translated by M.H. Shakir, Ansarian Publication, Qum, the Islamic Republic of Iran, 1993.
3. The Quran and the Last Prophet (in Persian), by ayatollah Makarem Shirazi, Dar AL-Kotob Al-Islamiah publishers, Qum, the Islamic Repeublic of Iran, 1996.
4. Cambridge International Dictionary of English, Cambridge University Press, 1995.
5. The Bible, The Quran and Science (Le Bible, le Coran et la Science)," The Holy Scriptures Examined in the Light of Modern Knowledge, by Dr. Maurice Bucaille, French Physician, Seghers, Paris, 1987, English version published by North American Trust Publication, 1978.

Aerospace in Holy Quran

- Aerodynamics and Flight Control
- Spacecraft
- Conquest of Space by Human

Aerodynamics and Flight Control

أَلَمْ يَرَوْا إِلَى الطَّيْرِ مُسَخَّرَاتٍ فِي جَوِّ السَّمَاءِ مَا يُمْسِكُهُنَّ إِلَّا اللَّهُ إِنَّ فِي ذَٰلِكَ لَآيَاتٍ لِقَوْمٍ يُؤْمِنُونَ ۝

Undoubtedly the flight procedure of the bird in nature has been an inspirational model for man to achieve his ancient aspiration of flight. For millennia, the idea of being able to fly occupied human dreams and fantasies. Waddling around on Earth's surface as majestic birds flew overhead perhaps human developed a form of wing envy. It took a long time for him to know the physical rules governing the nature by which he could overcome the obstacles and make the flight practicable in the beginning of the twentieth century.

The Wright brothers, Orville and Wilbur of Ohio are, of course, generally credited with being "first in flight" at Kitty Hawk, North Carolina. But this claim needs to be further delineated. Wilbur and Orville Wright were the first to fly a heavier-than-air, engine-powered vehicle that carried a human being - Orville, in this case - and that did not land at a lower elevation than its takeoff point. Previously, people had flown in balloon gondolas and in gliders and had executed controlled descents from cliff sides, but none of these efforts would make a bird jealous. Actually, Wilbur and Orville's first trip would not have turned bird heads either. At 10:35 A.M. Eastern Time on that historic day, December 17,

1903, the first of their four flights lasted twelve seconds, at an average speed of 6.8 miles per hour against a 30-mile-per-hour wind. The Wright Flyer, as it was called, had traveled 120 feet, a little more than the length of one wing on a Boeing 747 jumbo jet.

To make an object fly through the atmosphere, many principals and factors are involved: Aerodynamics, Mechanics of Flight, Mechanics of Material, Aero-elasticity, Viscous Fluid Flow, Principals of Propulsion, Dynamics, Statics, Mechanics of Fluids, Dynamics of Gases, etc. Among them, aerodynamics and mechanics of flight are the most significant principals by which the control and stability of a flight vehicle is ascertained. However, the Holy Quran states:

-Do they not see the birds, constrained in the middle of the sky? None withholds them but Allah; most surely there are signs in this for a people who believe[1].

Chapter16: verse79

This verse describes the birds as constrained during the process of flight. What does it really mean? Is it really true that they are constrained during flight? No kind of constraint is apparently seen around them. Nonetheless, from the standpoint of aerodynamics, any object (including a bird) which flies through the air is actually *constrained* by many different forces.

Some of the most known forces (termed constraints) are gravity, propulsive thrust (in case the flying object is an engine-powered vehicle), drag, lift and some environmental effects associated with atmospheric turbulence like winds etc. The resultant force determines the object's flight procedure.

For example, if the gravity overcomes the other forces, which are exerting the flight vehicle, then it would tend to fall on the ground. In case the thrust is dominant, depending on some other factors like the location of center of gravity, center of pressure, neutral point etc, the vehicle would have a dynamically and statically stable or unstable cruise flight.

Although, compressing any such account into a few pages is bound to lead to omission and over simplification, we will try our best to show you how these forces (or constraints) make a bird fly in the middle of the sky from the viewpoint of aerospace science. Amazingly, this like many other mysterious matters is contained in Holy Quran, which highlights the concepts of Aerodynamics, Control and Stability of a flying object.

Scientific Viewpoint

Aerodynamics

Aeronautics is defined as the science that treats of the operation of aircraft, also, the art or science of the operation of aircraft. Basically, with aeronautics, one is concerned with predicting and controlling the forces and moments of an aircraft that is traveling through the atmosphere. The term aerodynamics is generally used for problems arising from flight and other topics involving the flow of air. The dynamics of gases, especially of atmospheric interactions with moving objects.

Because the aerodynamic forces are dependent on the angular orientation of the airplane relative to its flight path, and because the resultant of them must exactly balance its weight, the equilibrium state is without rotation; that is, it is a motion of rectilinear translation.

Control and stability

A body is in equilibrium when it is at rest or in uniform Motion (i.e., has constant linear and angular moment). The most familiar examples of equilibrium are the static one; that is, bodies at rest. The equilibrium of an airplane in flight, however, is of the second kind; that is uniform motion.

A body in equilibrium may be in any one of three states--stable, unstable, or neutral equilibrium. It is in stable equilibrium if, when a torque is applied, the body tends to return to its original position. It is in unstable equilibrium if it continues to turn to a new position after the torque ceases to act. The body is in neutral equilibrium if it comes to rest wherever it may be when the torque is removed.

A body cannot be completely at rest unless all forces that tend to move the center of gravity in some direction and all torques are balanced. Once a complete balance exists, the body is said to be in equilibrium. In this case, the sum of all forces and torques acting on the body must equal zero.

Using this rule, scientists can establish equations for any number of forces acting on a body in equilibrium. They may then choose to study the relationship between the components of the balancing forces.

Stability or the lack of it is a property of an equilibrium state. The equilibrium is stable, if when the body is slightly disturbed in any of its degrees of freedom, it returns ultimately to its initial state.

The Gannet knows it without studying any books. Body cambered and tail down to reduce speed, increases lift, and the nose-down pitching moment. To trim, wings move forward to displace center of pressure ahead of center of gravity. Airflow begins to separate near wing trailing edges as speed is reduced, causing soft feathers to ruffle upwards, turbulating the boundary layer, delaying further separation and stall. Alula (bastard wing) acts as a slot by extending at the leading edge, inboard of the primary (pinion) feathers, to delay stall by keeping flow attacked. Landing gear coming down to increase drag, before rolling (wings half folded) into an evasive dive. (DMT Ettinger via Shell Aviation News, UK)

Quranic viewpoint

-Do they not see the birds, constrained in the middle of the sky? None withholds them but Allah; most surely there are signs in this for a people who believe.

Chapter16: verse79

In the preceding verse, the word *"Musakharat"* is the plural form of the past participle *"Musakhar"* meaning "conquered" or "constrained thing". This conquest of birds refers to a thorough concept in aeronautics, i.e. aerodynamics. It determines how a flying object is constrained by different forces during flight through the air.

The airplane is in a steady climb, meaning that the airplane is not accelerating. The vector sum of all of the aerodynamic and gravitational forces and moments on the airplane are equal to zero; that is, the airplane is in equilibrium.

In following figure, **V** represents the Velocity of the airplane's center of gravity. The thrust, **T** is the propelling force that balances mainly the aerodynamic drag on the airplane. T can be produced by a propeller, a turbojet, or a rocket engine. The total lift on the airplane is the sum of the lifts on the various components such as the wing, the tail, fuselage, nacelle, and propellers. In the level flight, the lift, **L** is mainly the vertical force upward on the wing.

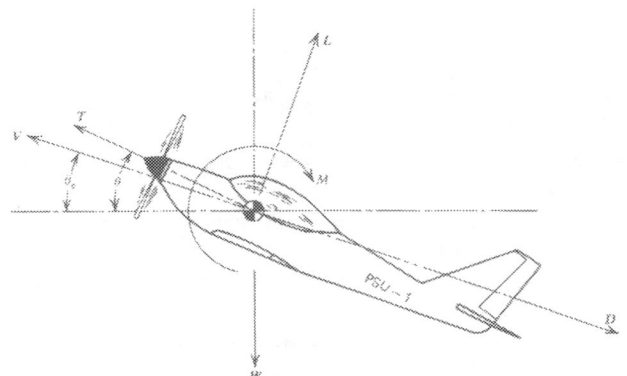

Forces and moments on an airplane in a steady climb.

Similar to the lift, the drag, **D** is defined as the component of all aerodynamic forces generated by the airplane in the direction opposite to the velocity vector, V.

W is the gross weight of the airplane and, by definition, acts at the center of gravity of the airplane and is directed vertically downloaded.

The pitching moment, **M** is defined as positive in the nose-up direction (clockwise) and results from the distribution of aerodynamic forces on the wing, tail, fuselage, engine nacelles, and other surface exposed to the flow. Obviously, if the airplane is in trim, the sum of these moments about the center of gravity must be zero.

What a miracle that this verse implies the aerodynamic forces, which constrain the bird during its flight process! By presenting such an inspirational account, the Holy Quran goes indeed, far beyond the level of human's knowledge in the 7th century. Imagine a man living 1400 years ago, pointing to a flying bird in the sky and telling the people: This bird, which is flying in the sky, is constrained! No one could easily understand him, as the flight itself has been a mythical paradigm of freedom since the incipient times. However, we now know that this is a true way of putting the matter.

Whatever Holy Quran implies, is deserved to be in relation with all the preceding scientific concepts. But, how could Muhammed have made such a delicate statement about the procedure of flight. Was he educated in an aerospace university 1400 years ago? No, never he was! The only thing, which remains to say, is that Holy Quran must be a revelation from God, Almighty!

References
1. Holy Quran translated by M.H. Shakir, by Ansarian Publication, Qum, the Islamic Republic of Iran, 1993.

Scientific Sources
- Aerodynamics, Aeronautics and Flight Mechanics by McCormick John Wiley and Sons Inc. 1995.
- Dynamics of Flight, Stability and Control, by Bernard Etkin, John Wiley and Sons Inc. 1996.
- Verne C. Cutler, Professor of Engineering Mechanics, University of Wisconsin, Milwaukee.
- The Design of the Aeroplane, by Darrol Stinton, BSP Professional books, UK, 1989.
- Fundamentals of Aerodynamics, by John D. Anderson Jr., Professor of Aerospace Engineering University of Maryland, McGraw-Hill, Inc. 1991.
- Man's Dream to Fly, http//: www.findarticle.com/.

Spacecraft

بِمَعْشَرَ الْجِنِّ وَالْإِنْسِ إِنِ اسْتَطَعْتُمْ أَن تَنفُذُوا مِنْ أَقْطَارِ السَّمَوَاتِ وَالْأَرْضِ فَانفُذُوا ۚ لَا تَنفُذُونَ إِلَّا بِسُلْطَانٍ ۝

Undoubtedly, spacecraft is one of the most brilliant favors to our civilization, which was produced by humankind in the twentieth century. It is by the means of spacecraft that human could travel to space and land on the moon in the 60's. Interestingly enough, it was beyond the comprehension of most men to imagine the possibility of space traveling before that time. This is quite obvious in the novels being composed and the movies produced in the 50's and aforetime. However, that is only the Quran, which has strongly stated that human can travel to space with a device named as "Sultan":

-O assembly of Jinns [demons] and Men! If you can penetrate regions of the heavens and the earth, then penetrate them! You will not penetrate them save with a Power [Sultan][1].

Chapter55: verse33

Why does the Quran name the power by which human can penetrate the space as 'sultan' (dominant)? To better conceive the context, we present you a brief scientific explanation regarding spacecraft, gravity and the physical law, which relates to this issue.

Scientific Viewpoint

Launch Vehicles

Launch vehicles send satellites and other spacecraft into space. These vehicles must be far more powerful than other types of rockets, because they carry more cargo farther and faster than other rockets. To place an object into orbit around Earth, the launch vehicle must reach a velocity of about 30,000 km/h (about 18,500 mph). To escape Earth's gravitational pull entirely and head into deep space, these rockets must attain a velocity, called an escape velocity, of about 40,000 km/h (about 25,000 mph). Engineers have found that the most efficient way for launch vehicles to reach these speeds is to use staged rockets, or rockets divided into different stages, one atop another.

Although some launch vehicles consist of just a single rocket, many are composed of a series of individual rockets, or stages, stacked atop one another. Such multistage launch vehicles are used especially for heavier payloads. With a multistage rocket, each stage fires for a period of time and then falls away when its fuel supply is used up. This lightens the load carried by the remaining stages. In some liquid-fuel boosters, strap-on solid-fuel rockets are used to provide extra thrust during the initial portion of ascent. For example, the Titan III booster has two liquid-fuel core stages and two strap-on solid-fuel motors. The largest example of a successful multistage booster was the Saturn V Moon rocket, which had three liquid-fuel stages and measured 111 m (363 ft), including the Apollo spacecraft, in length.

Despite their utility, most multistage boosters are not reusable, which makes them expensive. Cost-conscious engineers have focused on creating a single-stage-to-orbit (SSTO) vehicle. In an SSTO, the entire spacecraft and booster would be integrated into one fully reusable unit. If successful, this approach would reduce the costs of reaching Earth orbit. However, the technical challenge is enormous: A full 89 percent of an SSTO's total weight must be reserved for fuel, a much higher proportion than any previous launch vehicle. The payload, the crew, and the weight of the vehicle itself must make up only 11 percent of the SSTO's total weight.

Thruster

Many spacecraft use small rockets called thrusters to move around in space. Thrusters can change the speed and direction of a spacecraft. They allow a spacecraft to steer in space, to jump to a higher orbit, or to fall back to Earth.

How Rockets Work

All rockets-whether small or large, simple or complex-work by the basic principle of action and reaction, which was formulated by English scientist Sir Isaac Newton in 1687. Newton's third law of motion states, ì For every action there is an equal and opposite reaction.î In the case of the rocket, the expulsion of exhaust gases from the rear is the action, and the forward movement of the rocket is the reaction.

A. Action and Reaction

The motion of a rocket is much like the motion of a balloon losing air. When the balloon is sealed, the air inside pushes on the entire interior surface of the balloon with equal force. If there is an opening in the balloon's surface, the air pressure becomes unbalanced, and the escaping air becomes a backward movement balanced by the forward movement of the balloon.
Rockets produce the force that moves them forward by burning their fuel inside a chamber in the rocket and then expelling the hot exhaust that results.

Rockets carry their own fuel and the oxygen used for burning their fuel. In liquid-fueled rockets, the fuel and oxygen-bearing substance (called the *oxidizer*) are in separate compartments. The fuel is mixed with the oxygen and ignited inside a combustion chamber. The rocket, like the balloon, has an opening called a nozzle from which the exhaust gases exit. A rocket nozzle is a cup-shaped device that flares out smoothly like a funnel inside the end of the rocket. The nozzle directs the rocket exhaust and causes it to come out faster, increasing the thrust and efficiency of the rocket.

Some early scientists believed that rocket exhaust needed something to push against (such as the ground or the air) in order to move the rocket. Rockets traveling in the vacuum of space, however, demonstrated that this belief was not true. In fact, rockets produce more thrust in the vacuum of space than on Earth. Air pressure and friction with the air reduce a rocket's thrust by about 10 percent on Earth as compared to the rocket's performance in space.

B. Thrust and Efficiency

Thrust is a measurement of the force of a rocket, or the amount of ì pushî exerted backward to move a rocket forward. Thrusts vary greatly from rocket to rocket. Engineers measure thrust in units of weight or force

(Newtons [N] in the metric system and pounds [lb] in English measurements).

Staging

Rockets are very powerful, but it is often more efficient to use several rockets, rather than a single rocket, to move an object to the desired place. Launch vehicles often use more than one rocket engine, or stage, during a mission. In rockets that use stages, the stages are stacked on top of each other. The stage on the bottom of the stack is the first one to fire.

In some rockets that use stages, the first stage has additional rockets attached to the outside, acting as *boosters* to further increase the thrust. Rockets can theoretically use any number of stages, but the complications caused by coordinating the firing times of the stages make it impractical to have too many. The huge Saturn V rocket that sent Apollo astronauts to the Moon had four stages, including the Apollo spacecraft's own rocket.

The first and most powerful stage lifts the launch vehicle into the upper atmosphere. The first stage then separates from the rest of the rocket and falls toward Earth. Some first stages, such as the space shuttle's booster rockets, can be recovered. Others, such as the first stage of the huge Saturn V Moon rocket, burn up in the atmosphere once their fuel is expelled and they drop off the launch vehicle.

The second stage carries less weight than the first stage, because the first stage has dropped off of the rocket. When the second stage takes over, the vehicle reaches a much higher speed; the second stage, however, also uses up its fuel and drops off. The third stage fires and places the spacecraft into orbit (for a mission designed to orbit Earth).

On deep space missions, the third stage allows the spacecraft to reach escape velocity and head away from Earth. For some missions, three stages are not adequate.

Engineers can reduce the number of stages a launch vehicle needs by getting a rocket closer to its destination through some other means. For example, an airplane carries the Pegasus rocket, which sends spacecraft into space, to a high altitude first. The rocket then fires and carries its cargo into orbit.

Rocket Flight

Rockets are used for many different applications, but they share some aspects of their flight profiles (the actions and the order of the actions that they perform during flight). All rockets require some structure or method with which they can be launched. They also require a design that provides stability and control in flight.

Stability and Control

Rockets need some means of stabilization to help them fly in an even flight path through changes in air pressure, wind, uneven burning of the propellant, or slight irregularities in the balance of the rocket itself. Engineers apply Sir Isaac Newton's first law of motion to control rocket stability. This law states that an object in motion tends to stay in motion. The specific case of this law used in rocket stability is that a spinning body tends to keep spinning in the same orientation. One application of this law is to make the rocket spin. A spinning rocket is resistant to directional changes, making its flight more stable. Spin-stabilized rockets use special fins, or vanes, in the path of their exhaust. The vanes are oriented so that the rocket spins as the hot exhaust passes over the vanes.

Gravity

When you jump in the air or throw a ball up, you and the ball always come back down to the ground again. This is because an invisible force called gravity pulls everything towards the center of the earth. Gravity is a force that pulls two things together, but it is only strong enough to be noticed in large things like the earth, the sun or the moon. It keeps people, animals, buildings and trees firmly in place and prevents the air from drifting off into space.

The law of universal Gravitation states that every particle of matter in the universe attracts every other particle with a force directly proportional to the products of their masses and inversely proportional to the square of the distance between them. Consequently, the gravitational pull exerted by the earth upon all other bodies (including spacecraft) diminishes with distance from the earth.

However, it also makes it very difficult to escape from the earth if you want to go into space. To do this you must travel very fast: 40000km per hour (25000mph), about 20 times as fast as Concorde. This is called the escape velocity and the speed you need to escape from the gravity of the earth. Moons and small planets have less gravity, and so a lower escape velocity, than large planets or the sun. If you do not want to escape completely but simply to circle the earth, you only have to go about 12 times as fast as Concorde.

The only engines that can give a spacecraft this speed and still work in space are rockets. The jet engines used in aircraft cannot work in space because they take oxygen from the air to make their fuel burn. Rockets can work in space, where there is no air, because they carry their own supply of oxygen with them, sometimes in liquid form.

A rocket works in a very simple way. Its performance is based on the Newton's third law: To every action there is always an equal and opposite reaction; or, the mutual actions of two bodies upon each other are always equal, and act in opposite directions.

The fuel burning in the engine makes hot gases. These rush out of the engine nozzle, pushing the rocket forward according to this law. If you blow up a balloon and let it go, it shoots forward like a rocket so the air rushes out[2].

Aerodynamic forces on the lifting surfaces (for example, the wings) of an aircraft keep it aloft against the force of gravity, but a space vehicle cannot stay aloft in this way because of the absence of air in space. The spacecraft, therefore, must orbit if it is to remain in space. Aircraft flying in the earth's atmosphere can use propellers and winged surfaces for propulsion and maneuvering, but spacecraft cannot do so because of the lack of air. A space vehicle must rely on the reaction of rockets for propulsion and maneuvers, based on Newton's laws of motion. When a spacecraft fires a rocket blast in one direction, reaction against the rocket exhaust imparts momentum to the spacecraft in the opposite direction.

The gravitational field, however, extends to an infinite distance; gravity does not cease to act at any altitude. A spacecraft is said to be weightless when it is in orbit around the earth (or around any other celestial body) because the centrifugal effect (which acts away from the center) is then equal and opposite to the force of gravity. Under these conditions, objects in a spacecraft seem to float in space. In the same way, the moon does not fall toward the earth because of the centrifugal effect that balances the force of gravity[3].

Quranic Viewpoint

An amazing idea can be deduced from the word "Sultan" in the mentioned verse, i.e. spacecraft. The word "Sultan" is derived from the verb "salata" (To dominate) a subject noun, which means "dominant" in Arabic. Thus, the verse says: "You can not penetrate the space but with Sultan (dominant)". Namely, you can not penetrate the diameters of the heavens and the earth unless with something, which dominates or overcomes. Overcome to what? We all know the answer: "The gravity".

The fact of this matter is that this verse explicitly points out to the necessity of applying a device named Sultan (now called spacecraft), which has the *power* to overcome the force of gravity enabling humankind to travel (penetrate) to space.

One must note that the text of the Quran predicts not only penetration through the regions of the Heavens, but also the Earth, i.e. the exploration of its depths too.

How could all this knowledge come to illiterate Muhammed who lived 1400 years ago in the dessert of Arabia? Did he study in an aerospace university or a space agency in the desert peninsula of Arabia? Absolutely, it must have had a non-human source, which is a revelation from God, Almighty Himself.

References
1. Holy Quran translated by Arthur J. Arberry, Ansarian Publication, Qum, the Islamic Republic of Iran, 1993.
2. Space, Stars, Planets and Spacecraft, by Dorling Kindersley Publishers Limited, London, 1990.
3. 2000 Encarta, Microsoft Corporation.

Conquest of Space by Human

بِمَعْشَرَ الْجِنِّ وَالْإِنْسِ إِنِ اسْتَطَعْتُمْ أَن تَنفُذُوا مِنْ أَقْطَارِ السَّمَاوَاتِ وَالْأَرْضِ فَانفُذُوا ۚ لَا تَنفُذُونَ إِلَّا بِسُلْطَانٍ ۝

Ours is an era of space probes, computer chips, laser surgeries and cloning. Today, speaking about space exploration is indeed commonplace. We always hear and see on the media the space flights, sending probes to the far planets of the solar system and deep into space, shuttles, satellites, space stations, space walks, Space Telescopes, etc.

However, the Quran clearly predicted human's advancement and the possibility of space flight 1400 years ago and even clarified how human would make it happen:

-O assembly of Jinns and the men, if you are able to pass through the regions of the heavens and the earth, then pass through; You can not pass through but with authority (Power)[1].

Chapter55: verse33

The Quran through this glorious verse conspicuously, clarifies that humankind is able to penetrate the regions of space by a power (See "Spacecraft"). To speak about human's possibility of space flight in the seventh century is not a matter easily neglected. In the context of this time and the place surely by the unlettered person, who made this unbelievable statement, leads us to this fact that Holy Quran is not an ordinary book. We may feel compelled to believe in the preternaturalness aspect of the Quran. On reading it one is at once convinced that it is the word of God, for, no man can write such perfect guidance on so many subjects.

To have a better apprehension of this verse, first we have to offer you some scientific viewpoints. The following passage is a guideline for clarification.

Scientific Viewpoint

Science of Space Exploration

Space is a harsh environment for humans and human-made machines. Radiation from the Sun and other cosmic sources can weaken material and harm the human body even at the great distance we are from the sun on earth. In the vacuum of space, objects implode and can become boiling hot when exposed to the Sun. They can crystallize from the freezing cold when in the shadow of Earth or some other body. Scientists, engineers, and designers must make spacecraft and protective wear for men that can withstand these extreme conditions and more.

Space Exploration quest to use space travel to discover the nature of the universe beyond Earth. Since ancient times, people have dreamed of leaving their home planet and exploring other worlds. In the later half of the 20th century, that dream became reality.

Human exploration and the conquest of space began with the launching of the Russian Satellite, Sputnik I on Oct. 4, 1957. Then followed the launching of humans into space, which eventually led to a manned space flight to the Moon on July 20th, 1967. Thus, the dream of humanity since the dawn of civilization to go beyond the earth to the heavens was realized.

A human first went into space in 1961. Since then, astronauts and cosmonauts have ventured into space for ever greater lengths of time, even living in orbit around the earth for months on end. Two dozen people have circled the Moon or walked on its surface. At the same time, robotic explorers have journeyed where humans could not go, visiting all but one of the solar system's major worlds. Unmanned spacecraft have also visited a host of minor bodies such as moons, comets and asteroids. These explorations have sparked the advance of new technologies, from rockets to communication equipment to computers. Spacecraft studies have yielded a bounty of scientific discoveries about the solar system the Milky Way Galaxy and the universe. In addition, they have given humanity a new perspective on Earth and its neighbors in space.

It was inevitable that humans would follow their unpiloted creations into space. Piloted space flight introduced a whole new set of difficulties, many of them concerned with keeping people alive in the hostile environment of space. In addition to the vacuum of space, which requires any piloted spacecraft to carry its own atmosphere, there are other deadly hazards: "Solar and cosmic radiation, micrometeorites (small bits of rock and dust) that might puncture a spacecraft hull or an astronaut's pressure suit, and extremes of temperature ranging from frigid darkness to broiling sunlight. It was not enough simply to keep people alive in space-astronauts needed to have a means of accomplishing useful work while they were there".

It was necessary to develop tools and techniques for space navigation, and for conducting scientific observations and experiments. Astronauts need to be protected when they ventured outside the safety of their pressurized spacecraft to work in the vacuum. Missions and hardware would be carefully designed to help ensure the safety of space crews in any foreseeable emergency, from liftoff to landing.

The challenges of conducting piloted space flights were great enough for missions that orbited Earth. They became even more daunting for the Apollo missions, which sent astronauts to the Moon. The achievement of sending astronauts to the lunar surface and back represents a summit of human space flight.

After the Apollo program, the emphasis in piloted missions shifted to long-duration space flight, as pioneered aboard Soviet and U.S. space stations. The development of reusable spacecraft became another goal, giving rise to the U.S. space shuttle fleet. Today, efforts focus on keeping people healthy during space missions lasting a year or more-the duration needed to reach nearby planets-and in lowering the cost of sending satellites into orbit.

Quranic Viewpoint

From this point of view, three verses of the Quran should command our full attention. One expresses, without any trace of ambiguity, what man should and will achieve in this field. In the other two, God refers for the sake of the unbelievers in Mecca to the surprise they would have if they were able to raise themselves up to the Heavens; He alludes to a hypothesis, which is not realized for the latter.

1) The first of these verses is Chapter 55, verse 33:

-O assembly of Jinns and Men, if you can penetrate regions of the heavens and the earth, then penetrate them! You will not penetrate them save with a Power [Sultan].

There can be no doubt that this verse indicates the possibility men achieve what we today call *"the conquest of space"*. The translation given here needs some explanatory comments:

a) The word 'if' expresses in English a condition that is dependent upon a possibility and either an achievable or an unachievable hypothesis. Arabic is a language, which is able to introduce a nuance into the condition, which is much more explicit. There is one word to express the possibility (ida), another for the achievable hypothesis (in) and a third for the unachievable hypothesis expressed by the word (lau). The verse in question has it as an achievable hypothesis expressed by the word (in). The Quran therefore suggests the material possibility of a concrete realization. This subtle linguistic distinction formally rules out the purely mystic interpretation that some people have (quite wrongly) put on this verse.

b) God is addressing the spirits (jinn) and human beings, and not essentially allegorical figures.

c) 'To penetrate' is the translation of the verb "nafatha" followed by the preposition "min". The words "tanfothou, fanfothou and tanfothouna" are the derivatives of the verb "nafatha". According to Kazimirski's dictionary, the phrase means 'to pass right through and come out on the other side of a body' (e.g. an arrow that comes out on the other side). It therefore suggests a deep penetration and emergence at the other end into the regions in question.

According to the verse, God clarifies the possibility of man's travel into space as a penetration. A keen scientific point exists in this Quranic statement. The word "Penetration" implies a concept of motion with a resistance. That is why no other words like "ascend", "enter", "go up" etc are used instead of this verb, since in order to penetrate the space we have to achieve the escape velocity. Otherwise, nothing can get into space.

Getting into Space

One of the most difficult parts of any space voyage is the launch. During launch, the craft must attain sufficient speed (escape velocity) and altitude to reach earth orbit or to leave earth's gravity entirely and embark on a path between planets. Scientists sometimes find it helpful to think of earth's gravitational field as a deep well, with sides that are steepest near the planet's surface. The task of the launch vehicle or booster rocket is to climb out of this well.

Escape Velocity

Escape velocity is minimum initial velocity required for an object to escape the gravitational attraction of an astronomical body, and to continue traveling away from it without the use of propulsive machinery. The escape velocity is usually given in terms of the surface-launch velocity, disregarding aerodynamic friction[2].

Escape velocity decreases with altitude and is equal to the square root of 2 (or about 1.414) times the velocity necessary to maintain a circular orbit at the same altitude. At the surface of the Earth, if atmospheric resistance could be disregarded, escape velocity would be about 11.2 km (6.96 miles) per second. The velocity of escape from the less massive Moon is about 2.4 km per second at its surface.

2) The other two verses are taken from Chapter15, verses 14 and 15. God is speaking of the unbelievers in Mecca, as the context of this passage in the sura shows:

-Though We opened to them a gate in heaven, and still they mounted through it,
-Yet would they say, "Our eyes have been dazzled; nay, we are a people bewitched!"[3].

The above expresses astonishment at a remarkable spectacle, different from anything man could imagine. The conditional sentence is introduced here by the word "lau", which expresses a hypothesis that could never be realized as far as it concerned the people mentioned in these verses.

Therefore, we have two passages in the text of the Quran: one of them refers to what will one day become a reality thanks to the powers of intelligence and ingenuity God will give to man. The other describes an event that the unbelievers in Mecca will never witness, hence its character of a condition never to be realized. The event will however be seen by others, as intimated in the first verse quoted above. It describes the human reactions to the unexpected spectacle that travelers in space will see: their confused or dazzled sight as in drunkenness, the feeling of being bewitched...

This is exactly how astronauts have experienced this remarkable adventure since the first human space flight around the world in 1961. It is known in actual fact how once one is above the Earth's atmosphere, the

Heavens no longer have the azure appearance we see from Earth, which results from phenomena of absorption of the Sun's light into the layers of the atmosphere. The human observer in space above the Earth's atmosphere sees a black sky and the Earth seems to be surrounded by a halo of bluish color due to the same phenomena of absorption of light by the Earth's atmosphere. The Moon has no atmosphere, however, and therefore appears in its true colors against the black background of the sky. It is a completely new spectacle therefore that presents itself to men in space, and the photographs of this spectacle are well known to present-day man.

Here again, it is difficult not to be impressed, when comparing the text of the Quran to the data of modern science, by statements that simply cannot be ascribed to the thought of a man who lived more than fourteen centuries ago.

References
1. Holy Quran translated by M.H. Shakir, by Ansarian Publication, Qum, the Islamic Republic of Iran, 1993.
2. Contributed By: Fred Landis, MS., D.Sc, Professor of Mechanical Engineering, College of Engineering and Applied Science, University of Wisconsin-Milwaukee.
3. Holy Quran translated by Arthur J. Arberry, Ansarian Publication, Qum, the Islamic Republic of Iran, 1993.

Further References
- The Bible, The Quran and Science (Le Bible, le Coran et la Science)," The Holy Scriptures Examined in the Light of Modern Knowledge, by Dr. Maurice Bucaille, French Physician, Seghers, Paris, 1987, English version published by North American Trust Publication, 1978.
- Quran: A Teacher to Modern Scientists, by Sabeel Ahmed, Co-chairman of the Da'wa Committee and Board of Director at the Muslim Community Center, Illinois, July 1997 http://www.iol.ie/.

Physics in Holy Quran

- Universal Gravitation
- Gravity of Earth
- Universality of Pairing
- The Well-balanced Universe
- Origin of Iron
- Water and Its Individual Properties

Universal Gravitation

اللَّهُ الَّذِي رَفَعَ السَّمَاوَاتِ بِغَيْرِ عَمَدٍ تَرَوْنَهَا ثُمَّ اسْتَوَىٰ عَلَى الْعَرْشِ وَسَخَّرَ الشَّمْسَ وَالْقَمَرَ كُلٌّ يَجْرِي لِأَجَلٍ مُسَمًّى يُدَبِّرُ الْأَمْرَ يُفَصِّلُ الْآيَاتِ لَعَلَّكُم بِلِقَاءِ رَبِّكُمْ تُوقِنُونَ ۝

خَلَقَ السَّمَاوَاتِ بِغَيْرِ عَمَدٍ تَرَوْنَهَا وَأَلْقَىٰ فِي الْأَرْضِ رَوَاسِيَ أَن تَمِيدَ بِكُمْ وَبَثَّ فِيهَا مِن كُلِّ دَابَّةٍ وَأَنزَلْنَا مِنَ السَّمَاءِ مَاءً فَأَنبَتْنَا فِيهَا مِن كُلِّ زَوْجٍ كَرِيمٍ ۝

Today scientists describe gravitational forces that hold the celestial bodies apart from each other and prevent them from colliding with each other. However, this is what the Quran explicitly stated through these verses fourteen hundred years ago:

-Allah, it is who raised up the heavens without visible supports, then mounted the throne, and compelled the sun and the moon to be of service, each runneth unto an appointed term; He ordereth the course, He detaileth the revelations, that haply ye may be certain of the meeting with your lord.
Chapter13: verse2

-He hath created the heavens without supports that you can see[1].
Chapter31: verse10

In the verses mentioned above, the statements "without visible support" and "without support that you can see" should indeed attract our attention to this subject. In the following passage, we will clarify the scientific purport of these verses, which indicates the miraculous aspect of Holy Quran.

Having read the verses above and compared them with the new scientific discoveries, it really strikes a nerve in us. How did this knowledge come to Muhammed, a Bedouin illiterate man living in the peninsula of Arabia in the seventh century? How was this conveyed to the first readers of the Quran? God tells us in the Quran that He is the One Who raised the sky (Chapter55: verse7) and that he holds it back from falling on the earth (Chapter22: verse65). How exactly does God do this?

Scientific Viewpoint

Our universe is governed by four forces or interactions; gravity, the electromagnetic force, the strong nuclear force, and the weak interaction. These interactions are brought about by a group of particles called gauge bosons, which are exchanged between the particles that make up matter. Physicists are attempting to show how the four forces may, in fact, be derived from a single fundamental force.

A gauge boson is a general term for the four types of particle that transmit force. The gauge bosons that are thought to transmit each force are gluons (strong nuclear force); weakons (weak nuclear force); gravitons (gravitational force); and photons (electromagnetic force).

Gravity

The law of universal Gravitation states that every particle of matter in the universe attracts every other particle with a force directly proportional to the products of their masses and inversely proportional to the square of the distance between them. Gravity is the force that both holds galaxies together and causes a pin to drop. The larger the masses of two objects and the closer they are together, the stronger the gravitational pull between them will be.

Many scientists think that gravity is carried by particles called gravitons, but these have yet to be found in any experiment. (See "Gravity of Earth").

Electromagnetic Force

The electromagnetic force acts between all particles that have an electric charge, such as electrons.

Electromagnetic forces between the atoms and molecules of a solid object give the object its rigidity. The force is also responsible for the behavior of magnets and for producing light. The electromagnetic force is carried by particles called photons, which are also the particles from which light is made.

The Strong Nuclear Force

The strong nuclear force exists within the nuclear (core) of an atom. It holds together the atoms neutrons and the positive charged protons. Protons are constantly trying to push away from each other and would fly apart were it not for the strong force). The strong nuclear force which holds them together is carried by particles called gluons.

The Weak Interaction

The weak interaction causes a type of radioactive decay (break up of the nucleus of an atom) called beta decay. Radioactive atoms are unstable because their nuclei contain too many neutrons. In the case of beta decay, a neutron changes into protons, giving off an electron (In this case called a beta particle). The weak interaction is carried by W and Z particles[2].

Quranic Viewpoint

Around fourteen hundred years ago, Holy Quran has explicitly announced the existence of invisible supports, which raise up the heavens.

In that time no one was cognizant of these supports, but now we know a lot about them. It is quite obvious that these invisible supports refer to the gravitational and other invisible forces in the universe. These miraculous verses confirm that Holy Quran is a revelation from the wise, the almighty God!

Dr. Bucaille, the famous author of the book 'The Bible, the Quran and Science' comments in his book: "In the New American Bible, a picture is drawn to show how the authors of the Bible imagined the world to look like. In that picture, the sky "resembles an overturned bowl and supported by columns". (The New American Bible, St. Joseph's Medium Size Edition, pp. 4-5). The earth in that picture is flat, and supported by pillars. After describing the picture at length, the editors of that Bible conclude by calling that idea of the world a "pre-scientific concept of the universe."

At the time when the Quran was being revealed, anyone could have easily believed this description, which was previously presented in the Bible. It is only in modern times that people would know better. How did the author of the Quran avoid this mistake?

God says in the Quran that He created the heavens "without any pillars that you can see". Again, the Quran says: "God is the One Who raised the heavens without any pillars that you can see". Dr. Maurice Bucaille comments: "These two verses refute the belief that the vault of the heavens was held up by pillars, the only things preventing the former from crushing the earth".

To be able to avoid that pre-scientific error, the author of the Quran must have been either a modern scientist, or God Himself[3]. We might feel compelled to admit the divinity of the Quran.

References
1. Holy Quran translated by Marmaduke Pickthall, George Allen and Unwin Ltd., London, Fifth Edition 1969.
2. Eyewitness Encyclopedia of Space and the Universe, by Dorling Kindersley Publishers Limited, London, 1990.
3. The Bible, The Quran and Science (Le Bible, le Coran et la Science)," The Holy Scriptures Examined in the Light of Modern Knowledge, by Dr. Maurice Bucaille, French Physician, Seghers, Paris, 1987, English version published by North American Trust Publication, 1978.

Gravity of Earth

أَلَمْ نَجْعَلِ ٱلْأَرْضَ كِفَاتًا ۝

Undoubtedly, everyone before Sir Isaac Newton (1642-1727) had felt the existence of a mysterious force inside the earth, which draws everything to itself. Apparently, no one had a clue before him about the generality of this law throughout the universe, among the heavens, planets, stars, giant galaxies and even between two small particles.

Absolutely, Newton in the 17th century was the first man who perceived this great law, by being inspired of dropping an apple. The discovery of this law was so important that some scientists have named the 17th century as "the century of Newton". He proved that every two bodies attract each other in direct ratio of masses and reversed ratio of squared distances"[1].

Does the Quran have anything to say about the existence of this phenomenon, i.e. the tendency of earth to draw everything to itself as the law of gravity? To find the answer, we should consider the following verse:

-Have We not made the earth to draw together to itself?
Chapter77: verse25

We suppose that this astounding Quranic description of this phenomenon is so vivid for readers that there is seemingly no need for any particular scientific explanation. However, to obtain a background on the concept of gravity and its relation with the mentioned glorious verse, we offer you an overview of what Newton presented as the "laws of Mechanics", which have been dominant till the present time, as the "principles of Mechanics".

Scientific Viewpoint

Gravitation is the mutual attraction of all masses in the universe. While its effect decreases in proportion to distance squared, it nonetheless applies, to some extent, regardless of the sizes of the masses or their distance apart.

The concepts associated with planetary motions developed by Johannes Kepler (1571-1630) describe the positions and motions of objects in our solar system. Sir Isaac Newton (1643-1727) later explained why Kepler's laws worked, by showing they depend on gravitation. Albert Einstein (1879-1955) posed an explanation of how gravitation works in his general theory of relativity.

In our solar system, planetary motions are orbits gravitationally bound to a star. Since orbits are ellipses, a review of ellipses follows[3].

Newton's Principles of Mechanics

Classical mechanics is governed by three basic principles, which were first formulated in the 17th and 18th centuries by Isaac Newton. He realized that the force that makes apples fall to the ground is the same force that makes the planets "fall" around the sun. Newton challenged to address the question of why planets move as they do. He established that a force of attraction toward the sun becomes weaker in proportion to the square of the distance from the sun.

The first law describes a fundamental property of matter, called inertia, as follows: Every body remains in a state of rest or in a state of uniform motion (constant speed in a straight line) unless it is compelled by impressed forces to change that state.

Under this law a moving body is at rest, as far as its own inertia is concerned, as long as its motion continues at the same speed and in the same direction. Therefore, particles (or even worlds) of matter will keep flying through empty space forever, without being driven by any force, until something compels them to change their motion.

Newton's second law describes the manner in which a force compels a change of motion, at a rate of change called acceleration stated as follows: Change of motion is proportional to the impressed force and takes place in the direction of the straight line in which that force is impressed.

This law is often stated in a different manner: the net force acting on a body is equal to the product of the body's mass times the resulting acceleration. It can be stated much more simply as a formula, using letters for force, mass, and acceleration: $F=ma$. The wording of the law, however, makes clear how an impressed force acts. It simply compels a change in the body's motion-its speed or direction-toward the direction in which the force is acting.

Newton's third law is stated as follows: "Action and reaction are equal and opposite". This law is often expressed as "for every action there is an equal and opposite reaction." The law states a fact that can upset many calculations unless taken into account.

It explains, for example, the saying that a man cannot literally lift himself by his own bootstraps. As he pulls up on his bootstraps, the bootstraps pull down on him. Action and reaction are equal and opposite. A striking modern example of action and reaction is jet propulsion [4].

Quranic Viewpoint

The word "Kefatan" in the verse is derived from the word "Kafata" which in Arabic means: "(It) drew something to itself"[5], or "it held it back"[6]. As an adjective, "kefatan" describes the earth as having the feature of drawing everything toward itself. In the meantime, this word somehow implies the concept of the mutual attraction between the two things. This is why Shakir, the famous translator of the Quran, uses the word "together" in the translation to comment this concept.

All these implications about this phenomenon are all explicitly in accordance with scientific facts. However, 1000 years before Newton this verse had been revealed in the Quran:

-Have We not made the earth to draw together to itself?
Chapter77: verse25

This verse is indeed a wonderful miracle, which proves the divinity of Holy Quran as a revelation from God Almighty.

References
1. The Quran and the Last Prophet (in Persian language), by ayatollah Makarem Shirazi, Dar AL-Kotob Al-Islamiah, Qum, the Islamic Republic of Iran, 1996.
2. Holy Quran translated by M.H. Shakir, Ansarian publication, Qum,

the Islamic Republic of Iran, 1993.
3. http://www.ipl.nasa.com/.
4. Contributed by Verne C. Cutler, Professor of Engineering Mechanics, University of Wisconsin Milwaukee.
5. Monjid Al-Tollab (An Arabic-Arabic Encyclopedia), by Fu'aad Afraum Al-Bostaani.
6. A Dictionary of Modern Written Arabic, (Arabic-English) by Hans Wehr, McDonald and Evans Ltd., London.

Universality of Pairing

وَمِن كُلِّ شَىْءٍ خَلَقْنَا زَوْجَيْنِ لَعَلَّكُمْ تَذَكَّرُونَ ۝

سُبْحَٰنَ ٱلَّذِى خَلَقَ ٱلْأَزْوَٰجَ كُلَّهَا مِمَّا تُنۢبِتُ ٱلْأَرْضُ وَمِنْ أَنفُسِهِمْ وَمِمَّا لَا يَعْلَمُونَ ۝

Before the dawn of the modern era, humans conceived that only animal life was divided into two genders, male and female. Then the discovery was made that this phenomenon was present in plants and vegetation too. We have found this reality existing in every creation, animate as well as inanimate, though in different forms.

In electricity, these two genders can be classified as positive and negative. North and South Pole describe it in magnetism, electron and proton in atoms, matter and antimatter etc. Even bacteria could be positive or negative, while the truth is that this creation is made of pairs[1].

However, the Quran speaks of the universality of pairing in everything. The following glorious verses should indeed attract our attention in this regard:

-*And all things We have created by pairs, that haply ye may reflect!*
                                                    Chapter51: verse49

-*Glory be to Him, who created all the pairs of what the earth produces, and of themselves, and of what they know not*[2].

Chapter36: verse36

Given a great amount of thought about the import of these verses may lead us to this reality that how strongly a man 1400 years ago, ascribed the generality of pairing to all the existing things in the universe. To comprehend the magnificence of these miraculous verses, we offer you what scientists have to say about the pairing of all things as a general governing law in the universe.

Scientific Viewpoint

The most familiar of the elementary particles are proton, neutron and proton, which as we have seen, are the constituent particles of ordinary atoms. But each of these can exist as a separate entity, as in cosmic rays, in the solar wind, in laboratory particle accelerators, or, for that matter, in the ionized gases that make up the bulk of stars.

We have learned in the 20th century, however, that these are by no means all the particles that exist.

First, for each kind of particles, there is a corresponding antiparticle. If the particle carries a charge, its anti has the opposite charge. The anti electron is the positron, of the same mass as the electron, but positively charged. The antineutron, like the neutron has no charge, but interacts with other matter opposite to the way the neutron does. The antiproton has a negative charge.

Some particles- among them the photon and graviton- are their own antiparticles. For them the particles and antiparticles are identical. Whole atoms could exist of positrons, antiprotons and antineutrons.

These particles constitute what is called *antimatter*. Such atoms do not exist around here, because when a particle comes in contact with its antiparticle, the two annihilate, turning into energy. Antimatter in our world of ordinary matter, therefore, is highly unstable (in large doses, it would be mighty dangerous!), but individual antiparticles are found in cosmic rays and can be formed in the laboratory[3].

Quranic Viewpoint

The Quran describes the phenomenon in the most astounding manner:

-Glory be to Him, who created all the pairs of what the earth produces, and of themselves, and of what they know not.

Chapter36: verse36

The Quranic statement: *"...And of what they know not"* is as true today as it was when Holy Quran was revealed. Though, we have discovered that every created thing exists in pair, we have yet to discover many things that exist.

As you notice the verses above, it is astonishingly stated that all things have been created in pairs; namely, all the particles in the world exist with their own corresponding antiparticles, as we noticed above.

At this point it is essential to emphasize that these amazing statements prove the authenticity of Holy Quran as a revelation from Allah, the Glorified, since an illiterate man had not logically been able to make such an exact clarification in regard with the physics of the universe. Holy Quran would be no less than an everlasting miracle in the sight of those who possess right minds. Allah, the Almighty says in the Quran:

-[And] Such similitude coin We for mankind that haply they may reflect.

Chapter59: verse21

References

1. The Bible, The Quran and Science (Le Bible, le Coran et la Science)," The Holy Scriptures Examined in the Light of Modern Knowledge, by Dr. Maurice Bucaille, French Physician, Seghers, Paris, 1987, English version published by North American Trust Publication, 1978.
2. Holy Quran translated by Marmaduke Pickthall, George Allen and Unwin Ltd., London, Fifth Edition, 1969.
3. Exploration of the Universe, by Abell, Morrison, Wolff, Saunders College publishing, USA, 1987.

The Well-Balanced Universe

وَٱلْأَرْضَ مَدَدْنَٰهَا وَأَلْقَيْنَا فِيهَا رَوَٰسِىَ وَأَنۢبَتْنَا فِيهَا مِن كُلِّ شَىْءٍ مَّوْزُونٍ ۝

وَإِن مِّن شَىْءٍ إِلَّا عِندَنَا خَزَآئِنُهُۥ وَمَا نُنَزِّلُهُۥٓ إِلَّا بِقَدَرٍ مَّعْلُومٍ ۝

ٱللَّهُ يَعْلَمُ مَا تَحْمِلُ كُلُّ أُنثَىٰ وَمَا تَغِيضُ ٱلْأَرْحَامُ وَمَا تَزْدَادُ وَكُلُّ شَىْءٍ عِندَهُۥ بِمِقْدَارٍ ۝

ٱلشَّمْسُ وَٱلْقَمَرُ بِحُسْبَانٍ ۝

وَيَرْزُقْهُ مِنْ حَيْثُ لَا يَحْتَسِبُ وَمَن يَتَوَكَّلْ عَلَى ٱللَّهِ فَهُوَ حَسْبُهُۥٓ إِنَّ ٱللَّهَ بَٰلِغُ أَمْرِهِۦ قَدْ جَعَلَ ٱللَّهُ لِكُلِّ شَىْءٍ قَدْرًا ۝

خَلَقَ ٱلسَّمَٰوَٰتِ وَٱلْأَرْضَ بِٱلْحَقِّ وَصَوَّرَكُمْ فَأَحْسَنَ صُوَرَكُمْ وَإِلَيْهِ ٱلْمَصِيرُ ۝

The Quran through some verses states that every existing thing in the universe is well balanced and measured:

-And the Earth have We spread out and cast headlands upon it, and planted a bit of everything to grow it there (so it is) well balanced[1].

Chapter15: verse19

-And there is not a thing but with Us are the stores thereof. And We send it not down save in appointed measure².

Chapter15: verse21

-And everything with Him is measured³.

Chapter13: verse8

-The sun and the moon follow a reckoning⁴.

Chapter55: verse5

-God has appointed a measure for everything⁵.

Chapter65: verse3

-He created the heavens and the earth with the truth, And He shaped you and shaped you well; and unto Him is the homecoming⁶.

Chapter64: verse3

Interestingly enough, this is a fact, which scientists perceive it in the creation. Scientists support the theory that all parts of creation occur in perfect balance and in precise amounts, which supports living, as we know it. For example, the ratio of oxygen to nitrogen in the air is 21%. What if this ratio were 50%? All the combustible materials on earth as well as plants and forests would burn as soon as they received the first flame of sky lightning.

All the animals breathe oxygen and breathe out the carbon dioxide, which is urgent for the genesis of plants. If this interchange of oxygen and carbon dioxide between animals and plants did not occur in the nature, they would use up all the oxygen or all the carbon dioxide. Consequently, the plants would wither and animals would be condemned to die.

The sunlight reaches to us through a balanced distance between sun and earth. If this distance were further than what it is, everything would freeze. In addition, if it were less than what it is now, everything would burn out. 'We are very fortune that the sun is exactly the way it is. If it were different in almost any way, life would almost certainly never have developed on earth'.⁷

The intensity of sunlight maintains the perfect balance for our sustenance and existence. Scientists state that the growth of plants depends directly on sunlight. However, they require a definite amount of sunlight to flourish. If this intensity of sunlight changed, there would have been no possibility of plants' existence and growth on earth.

When reproduction of a single cell takes place we get double the matter from it. In addition, through the same input, the output depends on the genetic makeup. So how exactly it is balanced!

Scientists in all fields such as chemists and botanists believe that the existing materials in plants (without exception) occur in a measure according to a definite amount and a marvelous accuracy. Wonderfully, this is the way Quran speaks about the elements in plants and describes them as well-balanced.

When Einstein first studied the universe at large using the General Theory of Relativity he discovered that his equations predicted a universe, which was either expanding or contracting. The best astronomical observations at the time were used to challenge this concept. He then modified his equations to satisfy the observations. This modification corresponds to the assumption that the whole universe is permeated with a constant pressure (which in his case balanced the expansion yielding a steady universe). This universal pressure is called the cosmological constant.

If we perceive the world around us with a mystic vision and a logical mind, we certainly notice the precise reckoning and balance in all things. Have you ever happened to ask yourself about the creation of your body? Is it not really wonderful that we have eyes to see, ears to hear, hands to work, fingers to hold, teeth to chew, feet to walk and run, nose to smell, kidneys to purify the blood, brain to think and analyze etc? Why do these organs exist in our body? What would happen to our life if we omitted any of our body members and organs?

So, is not really the creation of our body in the best way it could be? If it is casual to be like this, then why does not any one dare to remove any of his body members, e.g. blind his eyes etc, unless he is insane? Alternatively, why does not any scientist venture to offer a better suggestion to improve the creation of human body? We may ponder this verse of the Quran:

-He created the heavens and the earth with the truth, and He shaped you and shaped you well; and unto Him is the homecoming.

Chapter64: verse3

The following scientific passage indicates the scientists' ideas concerning the existence of an intelligent designation in the well-balanced universe. Scientists see the world through this panorama. They believe that this world is fine-tuned to produce human life on earth as the fundamental constants of nature have made it quite suitable for us to live. Otherwise, there would not be any chance of our being here. This reminds us a verse of the Quran pointing to this principle:

-And [Allah] hath made of service unto you whatsoever is in the heavens and whatsoever is in the earth; it is all from Him. Lo! Herein verily are portents for people who reflect.

Chapter45: verse 13

Scientific Viewpoint

Cosmythology

Was the universe designed to produce us [8]?

Recent developments in modern cosmology seized upon research to provide scientific support for the notion of intelligent design to the universe. The latest version of the argument from design exists around the so-called Anthropic coincidences (Carter, 1974, 1983; Barrow and Tipler, 1986; Davies, 1982; Gribbin and Rees, 1989).

The Anthropic coincidences support the contention that the values of fundamental constants of nature are *incredibly fine-tuned* for the production of life-perhaps even human life. This fine-tuning is said to be far too unlikely to have been accidental and that the only reasonable conclusion is intelligent design, with human life as the intent. Holmes Rolston, for example, in *The Christian Century* called it "Shaken Atheism" (1986), Sharon Begley in *Newsweek* described it as "Science and the Sacred" (1994), while Robert Wright asked in *Time*, "What does science tell us about God?" (1992).

Tuned For Life

No doubt, the universe would look quite different with the tiniest variation of the basic constants of physics. A slight difference in the strength of gravity, the charge of the electron, or the mass of the neutron, life as we know it would not exist.

The human race could not have evolved in a universe with different constants. Those who promote the notion of intelligent design think they have found confirmation in the way the universe seems to be exquisitely *balanced* on the tip of a needle for the purpose, they argue, of producing us.

Yet, no theory including the currently highly successful Standard Model of elementary particles and forces, predicts the values of the fundamental constants of the universe. None is able to specify such basic facts about the universe as why the proton has the mass it does, or why the hydrogen atom has the size it does. In the Standard Model, the basic constants of the universe must still be inserted by hand.

The element-synthesizing processes in stars depended sensitively on the properties and abundance of deuterium and helium produced in the early universe. Deuterium would not exist if the neutron-proton mass difference were just slightly different from its actual value. The relative abundance of hydrogen and helium also depends strongly on this parameter.

The hydrogen-helium abundance also requires a delicate *balance* of the relative strengths of the gravitational and the weak nuclear interaction. Any minor increase in this force and the universe would be 100 percent hydrogen, since all the neutrons in the early universe would then have decayed. An even the slightest reduction in force and few neutrons would decay before being bound up with protons in helium nuclei where insufficient energy prevents their decay. In this case, all the proton utilization would change and lead to a universe that was 100 percent helium. Neither of these extremes would have allowed for the existence of stars and life based on chemistry.

This tightrope balance includes the electron, which is a requirement to produce the heavier elements. Because the electron mass is less than the neutron-proton mass difference, a free neutron can decay into a proton, electron and neutrino. If this were not the case, the neutron would be stable and most of the protons and electrons in the early universe would have combined to form neutrons, leaving little hydrogen to act as the main component and fuel of stars.

It is also rather convenient that the neutron is heavier than the proton, but not so much heavier that neutrons are restricted from bonding in nuclei. The evolution of life on earth thus depends critically on these relative force strengths and mass differences. With the slightest change of these values, the variety and diversity of the chemical elements would not exist.

Carbon appears to be the chemical element best suited to act as the building block for the type of complex molecular systems that develop lifelike qualities. Even today, new materials assembled from carbon atoms exhibit remarkable, unexpected properties, from superconductivity to ferromagnetism.

However, it is carbon chauvinism to assume that only carbon life is possible. We can imagine life based on silicon or other elements chemically similar to carbon, but these would still require cooking in stars. Hydrogen, helium, and lithium, which were synthesized in the big bang, are all chemically too simple to be assembled into diverse structures.

Still, the argument remains that if a universe were created with random values of the physical constants, a universe with no life would have almost certainly been the result. Of course, no one would then be around to talk about it and the fact is we are here and talking about it.

Quranic Viewpoint

Our world is a vast array of wonders and marvelous mysteries, which have astonished the enlightened human today. As mentioned before, many scientists believe that all these wonders are vivid indications of the fact that the creation is based on an extremely precise measurement and reckoning, which is originated from the will of a wise designer. Let us see what the Quran has stated about the organization of the Universe 1400 years ago:

-And the Earth have We spread out and cast headlands upon it, and planted a bit of everything to grow it there (so it is) well balanced.
Chapter15: verse19

-And there is not a thing but with Us are the stores thereof. And We send it not down save in appointed measure.
Chapter15: verse21

-And everything with Him is measured.
Chapter13: verse8

-The sun and the moon follow a reckoning.
Chapter55: verse5

-God has appointed a measure for everything.
Chapter65: verse3

-He created the heavens and the earth with the truth, And He shaped you and shaped you well; and unto Him is the homecoming.
Chapter64: verse3

Modern astronomers are aware that the stars and planets exist within ranges of precise distances from each other. Had it not been for this fact, collision between them would be inevitable. The author of the Quran was also aware of this. In the Quran we read:

-The sun and the moon follow a reckoning.
Chapter55: verse5

Again, we read:

-For you [God] subjected the sun and the moon, both diligently pursuing their courses.
Chapter14: verse33

The phrase 'diligently pursuing their courses' is a translation of the Arabic term daa'ib which here means 'to apply oneself to something with care in a perseverant, invariable manner, in accordance with set habits'. And that indeed is how the sun and moon behave.

Another verse in the Quran says:

- *the stars are in subjection to His command.*

Chapter16: verse12

Order in the universe is essential for its preservation. God, who subjected them to that order, knew about it before any scientist.

Whatever the Quran mentions in regard with the organization of the universe is important because "these references constitute a new fact of divine Revelation"[9]. The Quran deals with this matter in depth making it vastly unique compared to preceding scriptures.

Dr. Maurice Bucaille, the author of the famous book "The Bible, The Quran and Science" points out that those who say that Muhammed authored the Quran think that the Arabs were very knowledgeable in the field of Science, and Muhammed was of course one of them. However, this explanation is based on the incorrect assumption that the Arabs knew Science before the Quran was revealed.

As pointed out by Dr. Bucaille, the fact is that Science in Islamic countries came after the Quran, not before. "In any case", writes Dr. Bucaille, "the scientific knowledge of that great period would not have been sufficient for a human being to write some of the verses to be found in the Quran"[10].

Dr. Maurice Bucaille also points out the important fact that the Quran does not contain "the theories, prevalent at the time of the Revelation, which deal with the organization of the celestial world". Had the author of the Quran been human, he or she would have naturally included the ideas prevalent at the time. During the intervening history, many of those ideas were disproved and shown to be quite inaccurate. How did the author of the Quran know enough to exclude those ideas, unless the author is God himself?

References
1. Holy Quran translated by Thomas Ballantine Irving, Suhrawardi Research and Publication Center, Teheran, 1998.
2. Holy Quran translated by Marmaduke Pickthall, George Allen and Unwin Ltd., London, Fifth Edition 1969.
3. ibid.
4. Holy Quran translated by M.H. Shakir, Ansarian publication, Qum, the Islamic Republic of Iran, 1993.

5. Holy Quran translated by Arthur J. Arberry, Ansarian Publishers, Qum, the Islamic Republic of Iran, 1999.
6. ibid.
7. From "Information Summaries", National Aerospace and Space Administration, June 1991.
8. From Skeptic vol. 4, no. 2, 1996, pp. 36-40, by Dr. Victor J. Stenger, Professor of Physics and Astronomy at the University of Hawaii, http://www.skeptic.com/.
9. The Bible, The Quran and Science (Le Bible, le Coran et la Science)," The Holy Scriptures Examined in the Light of Modern Knowledge, by Dr. Maurice Bucaille, French Physician, Seghers, Paris, 1987, English version published by North American Trust Publication, 1978.
10. ibid.

The Origin of Iron

لَقَدْ أَرْسَلْنَا رُسُلَنَا بِالْبَيِّنَاتِ وَأَنزَلْنَا مَعَهُمُ الْكِتَابَ وَالْمِيزَانَ لِيَقُومَ النَّاسُ بِالْقِسْطِ وَأَنزَلْنَا الْحَدِيدَ فِيهِ بَأْسٌ شَدِيدٌ وَمَنَافِعُ لِلنَّاسِ وَلِيَعْلَمَ اللَّهُ مَن يَنصُرُهُ وَرُسُلَهُ بِالْغَيْبِ إِنَّ اللَّهَ قَوِيٌّ عَزِيزٌ ۞

Iron was known in prehistoric times and the most known metal for human today. Undoubtedly, iron played a key role in the development of civilization and no one can deny it. The human has used iron in most of his life aspects since the incipient eras. Indeed, life in many ways has always depended on the existence of iron. It was utilized in hunting tools, cooking utensils, war devices etc.

Usually, it was a new invention or a new technology, which allowed a nation to conquer an empire. Iron is much harder than bronze, so when the knowledge of how to make iron tools and weapons spread from the Middle East, it led to a new round of wars and invasions. At this time, about 3000 years ago, men learned to ride on horseback. They became expert warriors, able to fire arrows from a charging horse. Horse-riding nomads left the steppe, causing more migrations and wars.

The Christian Bible's first book Genesis says that Tubal-Cain, seven generations from Adam, was "an instructor of every artificer in brass and iron." Smelted iron artifacts have been identified from around 3000 BC. A remarkable iron pillar, dating to about 400 AD, remains standing today in Delhi, India. This solid pillar is wrought iron and about 7.5 m high by

40 cm in diameter. Corrosion to the pillar has been minimal despite its exposure to the weather since its erection.

The Ancient Finnish Myths about the Iron

In the Origin of the Iron, we are returned to the Beginning of all, to the primeval fountain in the center of the Cosmos. There the molten iron was originated in the breasts of three maidens, the Maidens of Iron. They stood on a Nameless Meadow, on a Borderless Field, which is in the dreamtime without names, places, and events. The maidens yielded milk, which transformed into molten iron and steel.

This same theme of maidens (or a goddess or a cow or a she-goat) at the fountain in the center of the primitive world (Paradise) is repeated in the myths all over the world, also in the Genesis. From the Fountain of Life there originated four rivers of water or milk, which often were depicted as four maidens. These rivers carried life, and the iron, all over the world.

According to this Finnish myth, when dealing with a wound made by an iron blade by his incantations the sorcerer makes it known to the victim and also to the weapon that he is aware of the origin of the iron and so is able to control[1].

What is the origin of Iron?

The matter of fact is that the main element of all the heavenly bodies including stars, red giants, super red giants, white dwarfs, supernova, interstellar galactic materials etc, are all made up of iron. (For more information, you can refer to the article ì Death of the Universeî).

The scientists believe that iron is an extraterrestrial element that was sent to earth and not formed therein. The energy of the early solar system was not sufficient to produce elemental Iron.

The core of our earth ì consists mainly of iron 90% and nickel along with silicon, sulfur and heavy elements. Recent evidence indicates the core may be as hot at 6,600íC or 12,000íF at it center, hotter than the surface of the sun! The tremendous pressure at this depth keeps the white-hot inner core solid. Isolated from the mantle by the liquid outer core, the inner core spins faster than the rest of the planet, gaining a full turn in about 300 years. This differential is the cause of the earthís magnetic field.î (Garrison, 1999, p. 54-55.) Essential during the formation of the earth the iron was placed deep within our Earth and the subsequent cooling formed the crust[2].

Scientists have only recently discovered the relevant facts about the iron's formation process. However, the verse 25 of chapter 57 ì Chapter of Ironî states that the iron is sent down from the heavens for the benefit of humankind:

-And We sent down iron, wherein is great might, and many uses for men[3].
Chapter57: verse25

To perceive the inspiration in this wonderful Quranic statement, the reference must be made to the recently discovered scientific facts.

Scientific Viewpoint

Iron is a type of atom, which was created in red giant stars. Iron atoms as stated above are believed to be common and of intense combinations at the center of the Earth. From the Anglo-Saxon word "iron" (the origin of the symbol Fe comes from the Latin word "*ferrum*"), which means, "iron".

Basic Information:
Name: Iron
Symbol: Fe
Atomic Number: 26
Atomic Mass: 55.845 amu
Melting Point: 1535.0 C (1808.15 K, 2795.0 F)
Boiling Point: 2750.0 C (3023.15 K, 4982.0 F)
Number of Protons/Electrons: 26
Number of Neutrons: 30
Classification: Transition Metal
Crystal Structure: Cubic
Density @ 293 K: 7.86 g/cm^3
Color: Silvery

Atomic Structure:
Number of Energy Levels: 4
First Energy Level: 2
Second Energy Level: 8
Third Energy Level: 14
Fourth Energy Level: 2

Facts:
Date of Discovery: Known to the ancients
Discoverer: Unknown

Name Origin: Latin
Symbol Origin: From the Latin word *ferrum* (iron)
Use: steel, hemoglobin (carries oxygen in blood)
Obtained From: iron ores

Isotopes:

Isotope	Half Life
Fe-	8.3 hours
Fe-54	Stable
Fe-55	2.7 years
Fe-56	Stable
Fe-57	Stable
Fe-58	Stable
Fe-59	54.5 days
Fe-60	1500000.0 years [4]

Quranic Viewpoint

The Quran informs us that Iron was actually descended from the sky and only came to form part of the Earth's makeup. After speaking of descent, this is to say that it is a principal component of a meteorite class known as *siderites* and is a minor constituent of the other two meteorite classes as science today explains in the link mentioned above.

It is also an interesting fact that this verse is in chapter 57, which represents the atomic weight (not the avg. atomic weight but a stable middle isotope). The verse, which mentions the Iron, is number 25 and the total number of verses of this chapter is 29 thus four (04) remaining verses to end this magnificent chapter.

Also the number of neutrons 30 - 04 = gives 26 which is the atomic number of iron and the number of its protons.

Sheik Abdul Majid Zendani, a professor of Islamic studies in king Abdulaziz University in Jeddah, Saudi Arabia says about his meeting with the professor Armstrong. Professor Armstrong works at the National Aeronautics and Space Administration (NASA), where he is a well-known scientist there. He was asked a number of questions about the Quranic verses dealing with the expertise in Astronomy and iron in particular and how it was formed. He explained how all the elements in the earth formed. He stated that the scientists have come only recently to discover the relevant facts about that formation process. He said that the energy of the early solar system was not sufficient to produce elemental Iron.

In calculating the energy required to form one atom of iron, it was

In calculating the energy required to form one atom of iron, it was found to be about four times as much as the energy of the entire solar system. In other words, the entire energy of the earth or the moon or the planet Mars or any other planet is not sufficient to form one new atom of iron, even the energy of the entire solar system is not sufficient for that. That is why Professor Armstrong said that the scientists believe that iron is extraterrestrial that was sent to earth and not formed therein. Sheik Abdul Majid Zendani says: We read to him the Quranic verse saying:

-And we sent down Iron, in which is Great might, as well as many benefits for mankind.

Chapter57: verse25

Each time Professor Armstrong related a scientific fact, it was mentioned to him the relevant verse, which he agreed with. It was then presented to him: 'You have seen and discovered for yourself the true nature of modern Astronomy by means of modern equipment, rockets, and space ships, developed by man. You have also seen how the Quran 14 centuries ago mentioned the same facts, so what is your opinion about these?

He replied: 'That is a difficult question, which I have been thinking about since our discussions here. I am impressed that how remarkably some of the ancient writings seem to correspond to modern and recent Astronomy'[5].

Surely the information contained within the Holy Quran baffles even the greatest minds of our time, because this truth transcends man's comprehension at the time it was revealed and surely its miracle is obvious even to nonbelievers.

References
1. The Iron, at http://www.geocities.com/.
2. Oceanography, Garrison, Tom, 3rd edition, Wadsworth Publishing, Belmont CA, 1999 or http: www.g3oceanography.com.
3. Holy Quran translated by Arthur J. Arberry, Ansarian Publication, Qum, the Islamic Republic of Iran, 1993.
4. Periodic Table: Iron, at http://www.chemicalelements.com/.
5. Formation of Iron, http://www.beconvinced.com/.

Water and Its Individual Properties

وَهُوَ ٱلَّذِى خَلَقَ ٱلسَّمَٰوَٰتِ وَٱلْأَرْضَ فِى سِتَّةِ أَيَّامٍ وَكَانَ عَرْشُهُۥ عَلَى ٱلْمَآءِ لِيَبْلُوَكُمْ أَيُّكُمْ أَحْسَنُ عَمَلًا ۗ وَلَئِن قُلْتَ إِنَّكُم مَّبْعُوثُونَ مِنۢ بَعْدِ ٱلْمَوْتِ لَيَقُولَنَّ ٱلَّذِينَ كَفَرُوٓا۟ إِنْ هَٰذَآ إِلَّا سِحْرٌ مُّبِينٌ ۝

وَٱللَّهُ خَلَقَ كُلَّ دَآبَّةٍ مِّن مَّآءٍ ۖ فَمِنْهُم مَّن يَمْشِى عَلَىٰ بَطْنِهِۦ وَمِنْهُم مَّن يَمْشِى عَلَىٰ رِجْلَيْنِ وَمِنْهُم مَّن يَمْشِى عَلَىٰٓ أَرْبَعٍ ۚ يَخْلُقُ ٱللَّهُ مَا يَشَآءُ ۚ إِنَّ ٱللَّهَ عَلَىٰ كُلِّ شَىْءٍ قَدِيرٌ ۝

أَوَلَمْ يَرَ ٱلَّذِينَ كَفَرُوٓا۟ أَنَّ ٱلسَّمَٰوَٰتِ وَٱلْأَرْضَ كَانَتَا رَتْقًا فَفَتَقْنَٰهُمَا ۖ وَجَعَلْنَا مِنَ ٱلْمَآءِ كُلَّ شَىْءٍ حَىٍّ ۖ أَفَلَا يُؤْمِنُونَ ۝

H_2O is a very familiar chemical formula to any high school student. We are so dependent on water that no one can deny its vital role in our life. Its importance in forming the creation in all of its aspects in general and the living-creatures in particular, as well as human kind, animals and plants is evident to all. However, the Quran mentions water's magnificence in forming all creation and making life in every living thing in the following passages:

-And We have made of water everything living, will not they believe?[1]

Chapter21: verse30

-And Allah has created from water every living creature[2].

Chapter24: verse45

-He it is Who created the heavens and the earth in six days and His throne was over the water[3].

Chapter11: verse7

By considering these inspirational verses of Holy Quran, the time of their revelation and also the newly discovered scientific facts regarding the water and its individual properties, it strikes a chord deep within us.

The fact that all life originated from water would not have been an easy thing to convince people of fourteen centuries ago. Indeed, if 1400 years ago you had stood in the desert and told someone, "All of this, you see (pointing to yourself), is made-up of mostly water, no one would have believed you. Proof of that was not available until the invention of the microscope. They had to wait to find out that cytoplasm, the basic substance of the cell, is made-up of 80% water. Nonetheless, the evidence did come, and once again the Quran stood the test of time.

We offer you the scientific findings about the properties of water and their strong correlation with the Quranic viewpoints. At the end, you also, like us, will certainly feel compelled to admit the authenticity of Holy Quran as a wonderful miracle from God Almighty, being revealed to Muhammed, the prophet of Islam.

Scientific Viewpoint

What Is Water?

Hydrogen and oxygen atoms comprise the basis of the water molecule. When two hydrogen and one oxygen atom bind together, water is formed. One molecule of water has two atoms of hydrogen (abbreviated H) and one atom of oxygen (abbreviated 0). The chemical formula for water is $H2O$.

Water covers almost three-quarters of earth's total surface-about 379 million square kilometers (146 million square miles). Water makes earth the "blue" planet; visually unique from all others in the solar system. Almost all plants, animals, and people need clean water to live a healthy life.

Water and Its Individual Properties / 163

Three phases to a quick-change artist

Changes in temperature can change water from a solid to a liquid to a gas and back again. At 0*C (32*F) pure water freezes. At 100*C (212*F) water boils. Andres Celsius (1701-1744) based his units of temperature (Celsius, abbreviated *C) on the three phases of water. Water becomes ice - a solid - at low temperatures. At medium temperatures, water stays in its liquid state. Water forms steam or vapor (gas at high temperatures).

A frozen pond warming in the winter sun may show all three phases at once. Solid ice floating on the lake surface, liquid water below and rising steam where the sun warms the melting ice.

Forming the link to life

All three phases of water - solid, liquid, and gas - form vital links to *life*. Liquid water accounts for two-thirds of our body weight. To stay healthy, we need about one liter (about one quart) of water each day. Water helps blood and its components transport oxygen and nutrients and remove waste products through our circulatory system. Each time we exhale, water vapor leaves our bodies.

It is the main substance, which is imperative for living creatures especially plants to survive.

Among the earth's varied environments, animals and plants adapt to water (or the lack of water) in different ways. Some plants and animals contain a great amount of water - a jellyfish is 95% water. In addition, watermelon aptly named contains 97% water. Most fishes and other aquatic creatures can only live when completed covered with water.

Some land animals, like amphibians, lay eggs in water. Others, like the desert kangaroo rat, seldom drank water but survive by eating seeds, plants, and metabolizing fat to produce water.

Green plants break down water (and carbon dioxide) during *photosynthesis* to produce oxygen and the simple sugars they require for energy.

Powerful Properties Shape Our World

When oxygen (0) and hydrogen (H) negative charge (OH-) atoms combine, they form a V-shaped, triangular molecule. While water molecules are electrically neutral, the oxygen atom holds a small negative charge and the two hydrogen atoms hold small positive charges. Scientists believe this unusual electrical balancing, called polarity, gives water some of its *remarkable properties.*

World's best dissolver

Scientists often call water the "universal solvent" because water can dissolve more substances than any other liquid. Why?

First, water molecules are very small and move easily around other atoms and molecules. Secondly, the negative charges on the oxygen atom and positive charges on the hydrogen atoms allow water molecules to interact with other molecules. Thirdly, water is very stable; at 2,000-C (3,632-F), only about 2% of water molecules break into parts. These parts are hydrogen ions with a positive charge (H+) and hydroxyl ions with a negative charge (OH-).

Some substances, like common table salt (sodium chloride), dissolve in water very easily. When placed in water, sodium chloride molecules fall apart. The positively charged sodium ion (Na+) binds to oxygen, while the negatively charged chloride ion (CI-) attaches to hydrogen. This makes a very stable "salty" water molecule.

Solid expansion

For most substances, solids are denser than liquids. But the special properties of water make it less dense as a solid - ice floats on water! Strong hydrogen bonds formed at freezing 0*C (32*F) lock water molecules away from each other. When ice melts, the structure collapses and molecules move closer together. Liquid water at 4*C (39.2*F) is about 9% denser than ice. This property plays an important role in lake and ocean ecosystems. Floating ice often isolates and protects animals and plants living in the water below.

A heated exchange

Pure water boils at 100'C (212'F), but extra energy is needed to push water molecules into the air. This is called *latent heat*-the heat required to change water from one phase to another.

Scientists have found one gram of water requires 2,500 *joules* of heat to change into gas at its boiling temperature. This extra energy is released when gas returns to liquid form.

It also takes a great deal of energy to raise the temperature of water from freezing (0*C or 32*F) to boiling (100*C or 212*F). This specific heat is the heat required to raise one gram of liquid water 1*C. One gram of water needs 4.18 joules of heat to warm 1*C. This is five times greater than the specific heat of sand. On a hot summer day, beach sand may quickly warm to the point that it's too hot to stand on while ocean water warms only a little.

Energy is also lost when water freezes. Water molecules release 334 joules of energy for every gram when moving from the high-energy phase of liquid water to the low-energy phase of ice. Nights when ice freezes often feel warmer than nights when ice melts.

Tension on top

Water molecules at the surface (next to air) are held closely together, forming an invisible film. Water's *surface tension* can hold weight that would normally sink. You can carefully float a sewing needle or paper clip on top of water in a glass.

Surface tension allows many aquatic insects, like water spiders and pod skaters, to "walk" across rivers and streams. Next to mercury, water has the highest surface tension of all commonly occurring liquids.

Sticky sides

The electrical attraction and surface tension of water molecules allow them to "hold on" or adhere to other substances such as glass, rocks, and soil. You can see this property when water creeps up the inside of a drinking glass. Water also clings to living things. Most plants have adapted to take advantage of water's *adhesion* that helps move water from the roots to the leaves.

One of the tallest plants is the redwood tree. Water moves its roots to its leaves, more than 95 m (310 ft.) above the ground. As a plant loses water through pores in the leaves, more water moves up from roots and stems to replace the lost water. The process of water loss by leaves is known as *transpiration.*

Global recycling

Water constantly moves between the earth and sky. When the sun warms the surface waters of lakes, streams, oceans, the water evaporates into vapor that rises upward in the sky. Plants and animals also lose water through leaves, sweat, excretion, or by exhaling.

As water vapor rises, it becomes cooler, losing energy and forming clouds. Water vapor condenses into rain or snow (precipitation). Rain falls from the sky, collecting in streams that flow into rivers that reach the sea. Surface water evaporates to rise as water vapor once again.

Scientists call this process the surface, less than *water cycle.* This global recycling of water creates weather (and water) in many forms- rain, snow, fog, floods, tornadoes, dew, clouds, sleet, frost, and hail.

Ocean Commotion

Oceans alone cover about 70% of the earth's surface. Winds blowing from high and low pressure regions in the tropics push surface water basins. Water currents traveling across the ocean meet continental landmasses, turning to flow along shorelines.

Generally, major water currents flow in United States flow in a clockwise direction in the Northern Hemisphere and in a counterclockwise direction in the Southern Hemisphere. The mightiest of all ocean currents, the Circumpolar Current, flows in a complete circle around Antarctica.

Sometimes major water currents change when winds weaken or blow in a different direction. One well-studied change is El Nino, the occurrence of warm water current off the coast of Ecuador and Peru. During some years, El Nino may raise water temperatures for 8,045 km (5,000 mi.) across the equatorial Pacific Ocean.

In addition to changing normal weather patterns, the warmer, nutrient-poor water affects animal life in the sea and on land. During the 1982-1983 El Nino event off South America, seabirds abandoned their nests, a fourth of the seal and sea lion population died, and the anchovy fisheries collapsed.

Every drop is precious. Even though water covers three-quarters of the earth's total surface, less than half of one percent is available fresh water. An estimated 97% is seawater, another 2% is locked in polar icecaps and glaciers, and the rest of the unavailable water is trapped deep below the earth's surface.

Available fresh water comes from many sources: surface rivers, streams, and lakes; underground water held in rock formations (aquifers); collected rainwater; and purified seawater[4].

Quranic Viewpoint

Considering these wonderful scientific facts about the life-giving properties of water in the nature clarifies the authenticity of this verse:

-*...And We have made of water everything living...*

This verse is one of the strongest proofs for Muhammed's prophecy. Since, none of these scientific facts about water were disclosed 1400 years ago, particularly to Muhammed as an illiterate man. Holy Quran simply exposed these facts to the atheists that they might be inclined to believe in Allah. Surely, this is considered to be the ultimate purpose of this verse, since it is remarked at the end: *Will not they believe?*

References
1. Holy Quran translated by M.H. Shakir, Ansarian Publication, Qum, the Islamic Republic of Iran, 1993.
2. ibid.
3. Holy Quran translated by Abdullah Yusuf Ali, Dar Al Arabia, Beirut, 1968.
4. Sea World, Inc/Sea World Education Department, 1996.

Meteorology in Holy Quran

- Protective Properties of Atmosphere
- Drop of Atmospheric Pressure
- Seven Layers of Atmosphere
- Reduction of Atmosphere
- Formation of Clouds and Rains

Protective Properties of Atmosphere

وَجَعَلْنَا ٱلسَّمَآءَ سَقْفًا مَّحْفُوظًا وَهُمْ عَنْ ءَايَـٰتِهَا مُعْرِضُونَ ﴿٣٢﴾

The Quran depicts the sky as a roof, which is withheld. Interestingly enough, this is the best description about the atmosphere and its protective properties from the viewpoint of science. In the following verse, two words should command our full attention; "roof and withheld":

-And We have made the sky a roof withheld, yet they turn away from its portents[1].

Chapter21: verse32

What is the general function of roof? It is obvious that the roof is used in order to cover the top of a building, vehicle etc, to protect us against the sun's severe light and heat, cold weather, wind, rain, hail, thunderstorm, lightening and withhold any thing harmful to our lives.

The Earth is constantly bombarded by meteoroids that disintegrate upon the atmosphere and by lethal rays emitted by the sun. This UV radiation is absorbed by the Ozone layer forming the outer fringe of our atmosphere. Thus, our atmosphere along with its Ozone layer is a protective covering for us. Life possibly could have not existed without it.

How could this precise simile have been presented by a Bedouin illiterate man in the seventh century? The logical answer is that this is no less than a miraculous revelation from Allah to remind us of His portents.

How meticulously, does our Lord sustain and protect us against this atmospheric destruction. Now, the scientific proof for the above description:

Scientific Viewpoint

Our planet's rapid spin and molten nickel-iron core give rise to an extensive magnetic field, which along with the atmosphere shielded us from really all of the harmful radiation coming from the sun and other stars. Earth atmosphere protects us from meteors as well; most of which burn up before they can strike the surface[2].

Ozone Layer

One of the important layers of atmosphere, which anyone may have heard about, is ozone. Ozone (abbreviated O3) is a heavy form of oxygen, having three atoms per molecule instead of the usual two. It has the property of being a good absorber of ultraviolet light. In absorbing the sun's short wave_ length ultraviolet light, the ozone is heated up and warms the parts of the atmosphere where it is present. Incidentally, the protective ozone layer helps to prevent some of the sun's dangerous ultraviolet radiation from penetrating to the earth surface[3].

Formation of the Ozone Layer

One billion years ago, early aquatic organisms called blue-green algae began using energy from the Sun to split molecules of H_2O and CO_2 and recombine them into organic compounds and molecular oxygen (O_2). This solar energy conversion process is known as photosynthesis. Some of the photosynthetically created oxygen combined with organic carbon to recreate CO_2 molecules. The remaining oxygen accumulated in the atmosphere, touching off a massive ecological disaster with respect to early existing anaerobic organisms. As oxygen in the atmosphere increased, CO_2 decreased.

High in the atmosphere, some oxygen (O_2) molecules absorbed energy from the Sun's ultraviolet (UV) rays and split to form single oxygen atoms. These atoms combined (27k JPEG) with remaining oxygen (O_2) to form ozone (O_3) molecules, which are very effective at absorbing UV rays. The thin layer of ozone that surrounds Earth acts as a shield, protecting the planet from irradiation by UV light. This thin layer of air also blocks out harmful rays from the sun and dust particles from space.

The amount of ozone required to shield Earth from biologically lethal UV radiation, wavelengths from 200 to 300 nanometers (nm) is believed to have been in existence 600 million years ago. At this time, the oxygen level was approximately 10% of its present atmospheric concentration. Prior to this period, life was restricted to the ocean. The presence of ozone enabled organisms to develop and live on the land. Ozone played a significant role in the evolution of life on Earth, and allows life, as we presently know it to exist[4].

Quranic Viewpoint

-And We have made the sky a roof withheld, yet they turn away from its portents.

Chapter21: verse32

These are the facts about our planet's atmosphere, which we have discovered in the few past decades. As you may notice the preceding verse, the Quran interestingly describes the sky (atmosphere) as a roof, which withholds and protects us from any kind of previously mentioned dangers. Needless to say, this would be considered as a miracle when being known that it is revealed about fourteen hundreds years ago, when no one knew any of these facts about the atmosphere and its precious protective properties.

Could life have existed without this withheld roof? Is it casually happened to be in the way it is now? Or it is originated from a merciful designer! The right-minded observer conceives the existence of a wonderful designation.

We have to feel compelled to bear witness that it is a great gift from God Almighty, the unique creator of the universe, who ordains us in the Quran not to be ignorant about his portents; the portents, which leads us to the righteousness of God. The signs, which help us to apprehend the authenticity of the Quran and that, it is the everlasting miracle from God, Almighty.

References

1. Holy Quran translated by Marmaduke Pickthall, George Allen and

Unwin Ltd., London, Fifth Edition 1969.
2. From "Information Summaries", National Aerospace and Space Administration (NASA), June 1991.
3. Exploration of the Universe by Saunders Publication.
4. Goddard DAAC, NASA official: Steve Kempler, GDAAC Manager, June28th, 2000.

Drop of Atmospheric Pressure

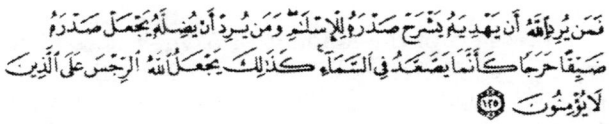

It has been proven by aeronautics that when man is exposed to high altitude, space; he suffers from some physiological symptoms that would vary between feeling tightness in his chest and getting into a critical stage if he continues being exposed to high altitude and low atmospheric pressure. Is there any relationship between the following verse and recent scientific discoveries?

-*Those whom God (in His plan) willeth to guide, He openeth their breast to Islam; [and] those whom He willeth to leave straying, He maketh their breast close and constricted, as if they had to climb up to the skies. Thus doth God (heap) the penalty on those who refuse to believe*[1].

Chapter6: verse125

The Holy Quran stated: "[and] those whom He willeth to leave straying, He maketh their breast close and constricted" mentions that tightness would occur in the chest when going up in the sky, which has now been proven by modern science.

Ascending to higher altitudes causes the symptoms of Hypoxia and Desparism. This leads to tightness of the chest because of the expansion of gases in the body cavities and their pressure on the lungs. In addition to this, the evolving of the dissolved gases (in the body) in the form of bubbles at high altitude causes severe chest pains.

So the words of Allah "He makes his breast closed" present a brief and precise explanation to the physiological changes occurring in man at these high altitudes.

This scientific fact mentioned in the Quran, was not discovered except after years of research. As for the word "constricted" it refers to what happens to man when he continues to rise in high altitude until he gets to the Critical Height which is 25,000 feet and higher when man loses consciousness due to failure of the nervous system. The pressure and density of the atmosphere decrease with height. At very high altitudes the heavier gases fail to rise until around 50 miles, hydrogen and helium are predominant.

Scientific Viewpoint

The pressure of the amount of oxygen needed to sustain life decreases rapidly with height until around 18000 ft the danger limit for the human pilot is reached and thereafter oxygen must be fed mechanically. Around 100000ft, there is no longer enough oxygen to support combustion in the most advanced turbojet engines now in service. Also mountain climbers ought to take extra oxygen with them while ascending to higher altitudes because of the shortage of oxygen there.

At higher and higher altitudes the air thins out more and more until it disappears into the extremely sparse gases of the magnetosphere at an altitude of several hundred kilometers [2].

Scientific facts concerning the state of man in high altitude

- The physiologically tolerated area for survival of human beings is from sea level up to 10,000 feet above sea level. The oxygen in this layer is physiologically sufficient for survival.
- The physiologically insufficient area is between 10,000 -50,000 feet. In this area there is a deficiency in oxygen, in addition to low atmospheric pressure. This would result in clear physiological symptoms on the human body, and so symptoms of Hypoxia (oxygen

deficiency) and Desparism (low atmospheric pressure) take place.
- Near space area (of the earth): from 50,000 feet. From the physiological point of view man can't live in altitudes higher 50,000 feet even if he breathes 100% oxygen. He should then wear a space suit to tolerate the decrease in atmospheric pressure and oxygen deficiency.

Symptoms of the stages of Hypoxia

Hypoxia is divided into four stages depending on the atmospheric pressure, altitude and blood oxygen concentration.
- From sea level to 10,000 feet no symptoms of Hypoxia appear.
- From 10,000-16,000 feet physiological compensatory systems of the human body prevent symptoms of Hypoxia from appearing except if the period of exposure is too long. Then breathing becomes faster and deeper, pulse and blood pressure also increase.
- From 16,000-25,000 feet, physiological systems do not function and cannot provide tissues with sufficient oxygen, and the previously mentioned symptoms appear. At this stage we find a clear explanation of the tightness of the chest one feels at this altitude.
- The Critical Stage is from 25,000 feet and higher. At this stage man completely loses consciousness due to the failure of the nervous system. Changes that occur in the chest reach their maxima at this altitude and then there will be complete physiological failure of heart functions and respiration.

Drop in Atmospheric Pressure

When man is exposed to low atmospheric pressure at high altitude (as what happens to air passengers when the pressure adjustment system fails inside the airplane), several symptoms occur as a result of the expansion of gases and their increase in the human body. Gases confined in the body cavities such as the stomach, when expanding press on the lungs, which cause breathing difficulty, disturbances, and tightness of the chest. The same thing would happen in the colon, lungs, teeth, middle ear and sinuses and all this causes severe pain in the body. In addition to this, all gases dissolved in the body cells e.g. nitrogen, would cause suffocation leading to severe chest pain[3].

That is why the astronauts wear spacesuits while they get out of shuttles or space stations. The air pressure inside the spacesuits equals to what is on the ground. So the breathing would be easy. The other reason for wearing the spacesuits in the vacuum environment is to prevent the burst of the body. Since the inner pressure of our bodies is exactly the same as the air pressure out, we do not feel any force on our skins. But the fact of reality is that a terrible weight of 1.03 kg exerts over each square centimeter of our bodies, which is reacted by the inner pressure. Without this pressure, the gases in our blood would bubble out and we would die[4].

Quranic Viewpoint

So, we can come to the conclusion that the statement "He makes his breast closed and constricted, as if he is climbing up to the sky" refers to this physical fact that air pressure decreases in high altitudes and makes the breathing hard. This is a scientific matter that we know it at the present time, but 1400 years ago no one was cognizant of this reality, especially, in the warm desert peninsula of Arabia! This knowledge of Quran could have not had a human source and is absolutely a scientific proof for the divinity of this Holy Revelation.

References
1. Holy Quran translated by Abdullah Yusuf Ali, Dar Al Arabia Publication, Beirut, Lebanon, 1968.
2. The Anatomy of the Aeroplane, by Darrol Stinton, BSP Professional Books, UK, 1987.
3. By: Dr. Salah El Maghribi.
4. Eyewitness Encyclopedia of Space and the Universe, by Dorling Kindersley Publishers Limited, London, 1990.

Seven Layers of Atmosphere

أَلَمْ تَرَوْا كَيْفَ خَلَقَ اللَّهُ سَبْعَ سَمَوَاتٍ طِبَاقًا ﴿١٥﴾

وَبَنَيْنَا فَوْقَكُمْ سَبْعًا شِدَادًا ﴿١٢﴾

وَلَقَدْ خَلَقْنَا فَوْقَكُمْ سَبْعَ طَرَائِقَ وَمَا كُنَّا عَنِ الْخَلْقِ غَافِلِينَ ﴿١٧﴾

Earth's Atmosphere

Atmosphere is defined as the mixture of gases surrounding any celestial object that has a gravitational field strong enough to prevent the gases from escaping, especially the gaseous envelope of Earth.

A blanket of air, called the atmosphere, surrounds the Earth. It reaches over 560 kilometers (348 miles) from the surface of the Earth, so we are only able to see what occurs fairly close to the ground. Early attempts at studying the nature of the atmosphere used clues from the weather, the beautiful multi-colored sunsets and sunrises, and the twinkling of stars. With the use of sensitive instruments from space, we are able to get a better view of the functioning of our atmosphere.

The atmosphere, solar energy, and our planet's magnetic fields support life on Earth. It absorbs the energy from the Sun, recycles water and other chemicals, and works with the electrical and magnetic forces to provide a moderate climate. The atmosphere also protects life from high-energy radiation and the frigid vacuum of space.

The principal constituents of the atmosphere of Earth are nitrogen (78 percent) and oxygen (21 percent). The atmospheric gases in the remaining 1 percent are argon (0.9 percent), carbon dioxide (0.03 percent), varying amounts of water vapor, and trace amounts of hydrogen, ozone, methane, carbon monoxide, helium, neon, krypton, and xenon. The atmosphere, weighing down upon the surface of the earth under the force of gravitation, exerts a pressure, which, at see level amounts to nearly 15 Lb.\in2.

The study of air samples shows that up to at least 88 km (55 mi) above sea level the composition of the atmosphere is substantially the same as at ground level. The continuous stirring produced by atmospheric currents counteracts the tendency of the heavier gases to settle below the lighter ones.

Atmospheric Layers

We live at the bottom of an ocean of air-several oceans, in fact, as the atmosphere can be divided into seven layers distinguished by temperature. It is astonishing that the Quran explicitly remarks that seven layers exist above us:

-Have you not regarded how God created seven heavens one upon another[1]?

Chapter71: verse15

-And we have built above you seven strong (heaven).[2]

Chapter78: verse12

-And we have created above you seven paths, and we are never unmindful of creation[3].

Chapter23: verse17

To provide a vivid background to the readers, the reference must be made to the latest scientific findings.

Scientific Viewpoint

Seven Layers of Atmosphere

Separated by temperature, the atmosphere contains seven different layers. There is an upper and a lower atmosphere. A rocket travels through the 7 layers of the atmosphere and then goes into space.

Troposphere

Most of the atmosphere is concentrated near the surface of the earth, within about the bottom 10km. That is where clouds form and airplanes fly. Called the *troposphere*, this region is characterized by convection currents produced as warm air, heated by the surface, rises and is replaced by descending currents of cooler air. The convection generates clouds and other manifestations of weather.

This layer contains about 75 % of the total mass of the atmosphere. It is also the layer where the majority of our weather occurs. Rain, snow and wind only take place in the troposphere. Maximum air temperature occurs near the Earth's surface in this layer. With increasing altitude air temperature drops rapidly until a temperature of -55 degrees Celsius is reached at the top of the troposphere.

Stratosphere

The stratosphere, directly above the troposphere, extends from 20 to 48 kilometers above the Earth's surface. Ninety-nine percent of "air" is located in the troposphere and stratosphere. Most of the stratosphere is cold and free of clouds, but in its upper part, the atmosphere rises again.

In the stratosphere temperature increases with altitude because ozone gas found in this layer absorbs ultraviolet sunlight creating heat energy. The stratopause separates the stratosphere from the next layer.

Ozone layer

In the lower atmosphere, ozone, a form of oxygen with three atoms in each molecule, is normally present in extremely low concentrations.

Ozone is primarily found in the atmosphere at varying concentrations between the altitudes of 10 to 50 kilometers. This layer of ozone is also called the *ozone layer*. Ozone protects life from the harmful effects of ultraviolet radiation. Without it life, as we know it, could not exist on Earth.

The hot layer is due to the presence of ozone in that level of the atmosphere. Ozone (abbreviated O3) is a heavy form of oxygen having 3 atoms per molecule instead of the usual two.

This layer is an excellent absorber of ultraviolet light and thus acts as a protective layer to all life on earth. The layer of atmosphere from 19 to 48 km (12 to 30 mi) up contains more ozone, produced by the action of ultraviolet radiation from the sun. Even in this layer, however, the percentage of ozone is only 0.001 by volume. Atmospheric disturbances and downdrafts carry varying amounts of this ozone to the surface of Earth. Human activity adds to ozone in the lower atmosphere, where it becomes a pollutant that can cause extensive crop damage.

The ozone layer became a subject of concern in the early 1970s, when it was found that chemicals known as chlorofluorocarbons (CFCs), or chlorofluoromethanes, used as refrigerants and propellants in aerosol dispensers, were rising into the atmosphere in large quantities. The concern centered on the possibility that these compounds, through the action of sunlight, could chemically attack and destroy stratospheric ozone, which protects Earth's surface from excessive ultraviolet radiation. As a result, industries in the United States, Europe, and Japan replaced chlorofluorocarbons in all but essential uses.

Mesosphere

The region of the stratosphere extending upward from the ozone layer is sometimes called the mesosphere. The chemicals are in an excited state, as they absorb energy from the Sun. From 65 to 80 km, the temperature drops to below -50 degrees C again. In the mesosphere, the atmosphere reaches its coldest temperatures (about -90 degrees Celsius) at a height of approximately 80 kilometers. The mesosphere, between 50 and 80 kilometers altitude, is where meteors burn up.

In the mesosphere, which extends to about 50 miles, temperature drops again to as low as -173 degrees F. Meteors, small pieces of matter drawn to the atmosphere by earth's gravity, become visible to the naked eye as they enter the mesosphere and are heated through friction caused by collisions with air molecules. These "falling stars" usually disintegrate before they reach the earth's surface. Spectacular meteor showers can be observed at certain times of the year when the earth's orbit, passes through a swarm of particles generated from the breakup of a comet.

The mesopause separates the stratosphere from the next layer. The troposphere, stratosphere and mesosphere make up what is called the lower atmosphere. Scientists call the regions of the stratosphere and the mesosphere, along with the stratopause and mesopause, middle atmosphere.

Ionosphere

In the upper atmosphere, molecules of oxygen and nitrogen break up into individual atoms of these elements.

Ultraviolet radiation from the sun ionizes many of these atoms. Because of the relatively high concentration of ions in the air above 60 km (40 mi.), this layer, extending to an altitude of 640 km (400 mi.), is called the ionosphere, which reflects long-wave radio, making it possible to receive A.M radio broadcasts over very large areas.

From investigations of the propagation and reflection of radio waves, it is known that beginning at an altitude of 60 km (40 mi.). Ultraviolet radiation, X rays and showers of electrons from the sun ionize several layers of the atmosphere, causing them to conduct electricity; these layers reflect radio waves of certain frequencies back to Earth.

How do we listen to a radio program originating in a city far from our own? The radio waves transmitted from one point on earth reflect off the ions and return to the surface - but because of the earth's curvature, when they return they are hundreds of miles away from the point of transmission. The waves once again bounce up to the ionosphere, return to the earth, and so on.

Thermosphere

At an altitude of about 90-km (55-mi.), temperatures begin to rise. The layer that begins at this altitude is called the thermosphere, because of the high temperatures reached in this layer (about 1200°C, or about 2200°F). The thermosphere is the hottest layer in the atmosphere as oxygen molecules absorb solar radiation.

The thermosphere extends to 400 miles and is characterized by large fluctuations of temperature (thermo means "heat"). At these heights there are relatively few molecules and heat retention should be low. However, within the thermosphere solar energy is absorbed and reradiates heat. At its upper limits the temperature reaches 441 degrees F. The thermosphere is where the space shuttle orbits.

Exosphere

The region beyond the thermosphere is called the exosphere, which extends to about 9,600 km (about 6,000 mi), the outer limit of the atmosphere.

The exosphere is the outermost layer of the atmosphere, and extends to 40,000 miles above the earth. It is here that molecules escape from the atmosphere without colliding with other molecules. Throughout the history of our planet most of the lighter molecules have escaped through the exosphere ("exo" means "out of"), while the heavier molecules, such as nitrogen and oxygen, have remained. The upper atmosphere can also be divided into regions characterized by exchanges of energy.

Quranic Viewpoint

As seen in the preceding passage, there are seven intensified layers of air above us including troposphere, stratosphere, ozone, mesosphere, ionosphere, thermosphere and exosphere, which are all imperative for earth's life support system.

These scientific findings that we know now about our world, all are newly discovered. But as you noticed in the above verses 1400 years ago, the Holy Quran miraculously revealed the fact that seven strong paths exist above us, while no one on earth was cognizant of the existence of these seven layers and their properties:

-And we have created above you seven paths, and we are never unmindful of creation.
Chapter23: verse17

Now, we can comprehend the majesty of Holy Quran as a miracle for the presence of this knowledge, which proves its divinity.

References
1. Holy Quran translated by Arthur J. Arberry, Ansarian Publishers, Qum, the Islamic Republic of Iran, 1999.
2. Holy Quran translated by Marmaduke Pickthall by George Allen and Unwin Ltd., London, Fifth Edition 1969.
3. ibid.

Scientific Sources
- The Anatomy of the Aeroplane, by Darrol Stinton, BSP Professional Books, UK, 1987.
- Exploration of the Universe, by George Abell, Hot, Rinehart and Winston Inc., USA, 1969.
- British science journal Nature, March 25th, 1999 reported by Massachusetts Institute of Technology (MIT).
- 1997 Cislunar Aerospace, Inc., CA.
- 1999-2000 Britannica.com, Inc.

Reduction
of Atmosphere

أَوَلَمْ يَرَوْا أَنَّا نَأْتِى الْأَرْضَ نَنقُصُهَا مِنْ أَطْرَافِهَا وَاللَّهُ يَحْكُمُ لَا مُعَقِّبَ لِحُكْمِهِ وَهُوَ سَرِيعُ الْحِسَابِ ۝

Scientists now conceive that the earth loses matters from its outer atmospheric layers especially the exosphere, while evolving around itself and orbiting in its elliptical path around the sun.
Two verses of the Quran indicate this phenomenon. One is as follows:

-See they not how We visit the land, reducing it of its outlying parts[1]?
Chapter13: verse41

Are the outlying parts of the earth really being reduced? Let us go over the pertinent scientific data in this regard.

Scientific Viewpoint

Historical Atmosphere

Earth is believed to have formed about 5 billion years ago. In the first 500 million years a dense atmosphere emerged from the vapor and gases that were expelled during degassing of the planet's interior. These gases may have consisted of hydrogen (H_2), water vapor, methane (CH_4), and carbon oxides. Prior to 3.5 billion years ago the atmosphere probably consisted of carbon dioxide (CO_2), carbon monoxide (CO), water (H_2O), nitrogen (N_2), and hydrogen.

The hydrosphere was formed 4 billion years ago from the condensation of water vapor, resulting in oceans of water in which sedimentation occurred.

The most important feature of the ancient environment was the absence of free oxygen. Evidence of such an anaerobic reducing atmosphere is hidden in early rock formations that contain many elements, such as iron and uranium, in their reduced states. Elements in this state are not found in the rocks of mid-Precambrian and younger ages, less than 3 billion years old [2].

Exosphere

The exosphere is the outermost layer of the atmosphere, where molecular densities are low and the probability of collisions between is very small. The Earth's exosphere begins about 500 km (300 miles) above the terrestrial surface and extends out through the magnetosphere and beyond to the interplanetary medium and extends to 40,000 miles above the earth.

Temperatures in the Earth's exosphere remain constant with altitude, averaging about 1500 K. The earth's exosphere contains the hydrogen geocorona and the Van Allen radiation belts.

It is here that molecules escape from the atmosphere without colliding with other molecules. Throughout the history of our planet most of the lighter molecules have escaped through the exosphere (exo means "out of"), while the heavier molecules, such as nitrogen and oxygen, have remained [3].

The base of the exosphere is called the critical level of escape because, in the absence of collisions, lighter, faster-moving atoms such as hydrogen and helium may attain velocities that allow them to escape the planet's gravitational field. Most molecules, however, have velocities considerably lower than the escape velocity, so their rate of escape to outer space is quite low [4].

Quranic Viewpoint

To analyze the respective Quranic purport, let us revise the verse again:

-See they not how We visit the land, reducing it of its outlying parts?
Chapter13: verse41

What does the "outlying parts of the land" refer to? The word "Atraafiha" in the verse means "the surroundings and sides of it" in which the pronoun 'it' refers to the earth. Obviously, the "outlying parts around the earth" indicate the atmosphere.

The escape of hydrogen from the outermost atmospheric layers around the earth, which brings about the reduction of atmosphere, is deserved to be mentioned in relation to this verse from the Quran, since it is perfectly in compliance with modern scientific data.

This is a scientific fact, which is astonishingly deduced from the mentioned verse. The presence of this knowledge in Quran manifests the miraculous aspects of this book from the viewpoint of science and proves its divinity as a word from God, which was revealed to Muhammed.

References
1. Holy Quran translated by Marmaduke Pickthall, George Allen and Unwin Ltd., London, Fifth Edition 1969.
2. Goddard DAAC, NASA official: Steve Kempler, GDAAC Manager, June28th, 2000.
3. 1997 Cislunar Aerospace, Inc.
4. http://www.Britannica.com/ Inc.

Formation of Clouds and Rains

أَلَمْ تَرَ أَنَّ ٱللَّهَ يُزْجِى سَحَابًا ثُمَّ يُؤَلِّفُ بَيْنَهُ ثُمَّ يَجْعَلُهُ رُكَامًا فَتَرَى ٱلْوَدْقَ يَخْرُجُ مِنْ خِلَـٰلِهِ وَيُنَزِّلُ مِنَ ٱلسَّمَاءِ مِن جِبَالٍ فِيهَا مِنْ بَرَدٍ فَيُصِيبُ بِهِ مَن يَشَاءُ وَيَصْرِفُهُ عَن مَّن يَشَاءُ يَكَادُ سَنَا بَرْقِهِ يَذْهَبُ بِٱلْأَبْصَـٰرِ ۝

The clouds tend to move closer together and are condensed by winds. Consequently, the different electrical charges between the clouds cause the formation of raindrops and thunder and lightning while getting close to each other. Without approaching, clouds could not produce rain, snow, hail etc and this is what the Quran has clarified through this verse:

-Hast thou not seen how Allah wafteth the clouds, then gathereth them, then maketh them layers, and then seest the rain come forth from between them; He sendeth down from the heaven mountains wherein is hail, and smiteth therewith whom He will, and averteth it from whom he will. The flashing of His lightning all but snatcheth away the sight[1].

Chapter24: verse43

The miraculous evidence of this verse is the statement "then He gathereth them". The gathering of clouds is a vivid expression and an exact description for inter-attraction of clouds carrying positive and negative charges. Undoubtedly, the clouds are electrically charged and this is the fact that Franklin discovered for the first time, in 1752. On the other hand, it is obvious that the positive and negative charges attract each other and two positive or two negative charges repel themselves. Let us glance over the procedure of cloud's formation from the scientific viewpoint:

Scientific Viewpoint

Why is the sky blue?

If you were to travel 20 miles or so above the Earth's surface, the sky would appear black. What happens during light's descent to Earth that makes the sky take on a wonderful azure hue?

"White" sunlight passes through our atmosphere, and molecules in the air, primarily nitrogen, are just the right size to scatter light from the blue end of the visible spectrum. The other colors travel to the ground with little interference.

The blue light is scattered from molecule to molecule in the sky, until the light seems to be coming from every direction.

And Clouds are White Because...?

The answer is that the water droplets that make up clouds are much larger than the molecules that scatter blue light. The clouds scatter and reflect all the visible colors of light that strike them. Hence, we have white clouds. But if the cloud is thick enough, light does not penetrate completely through the cloud, resulting in dark, heavy-looking cloud bottoms.

Why do clouds form?

Clouds are nothing more than water vapor that condenses and accretes into a visible form. The usual mechanism is for moisture-laden air near the Earth's surface to be raised higher into the atmosphere either by an encroaching air mass or the heat of the sun. As the air is lifted, the pressure drops and the air is subsequently cooled. The combination of the two causes water vapor to condense.

Formation of Clouds

Clouds are suspensions of liquid water drops that arise from the condensation of gaseous water vapor and subsequently disappear as they evaporate and the water is returned to the gaseous phase. As much of the liquid water that comprise the cloud has its origin from evaporation of water from the surface of the earth, the heat released to the air during condensation warms the cloud layer of the atmosphere. Were this process of transferring energy from the surface to the atmosphere at cloud level not present sensible heating of the layer of air near the ground would result in desert like heat everywhere. Because the water drops form and evaporate their life and the life of clouds is limited.

Lightning

In fine weather, the Earth carries a negative charge and positive charge in upper atmosphere (ionosphere). Potential difference of several hundred thousand volts can exist between surface and ionosphere. As charge builds up in atmosphere due to atmospheric motions, discharges (lightning) must occur to maintain a steady state. Discharges occur along ionized channels where ionization is the process by which atmospheric molecules (small ions) or other suspended particles (large ions) are electrically charged. Water drops in a field of drops in a cloud become charged and elongated.

When a cloud approaches given point on the Earth's surface and/or given stage in development, charge in cloud reverses - negative charge at cloud base replaced first by pockets of + charge and then by + charge reversing the original polarity of the cloud. Now have a thunderstorm. A reversal of charge used as a predictor. Gradients may reach 20,000V/m and may be 5000V/m at a distance of 5 km from cloud center.

Why Does It Rain?

Clouds form when the relative humidity approaches 100 percent, and are comprised of tiny cloud water droplets (remember the air is saturated). But these tiny droplets are far from raindrop status. It takes over one million cloud droplets to create one raindrop.

If the cloud was left alone in this scenario, it would take days to generate rain droplets through condensation. Consequently, rain would never form.

Scientists have studied cloud types and have realized that rain clouds are formed and shaped according to definite systems and certain steps connected with certain types of wind and clouds.

One kind of rain cloud is the cumulonimbus cloud associated with thunderstorms. Meteorologists have studied how cumulonimbus clouds are formed and how they produce rain, hail, and lightning. They have found that cumulonimbus cloud go through the following steps to produce rain:

(1) The clouds are pushed by the wind: Cumulonimbus clouds begin to form when wind pushes some small pieces of clouds (cumulus) to an area where these clouds converge.

2) Joining: Then the small clouds join together forming a larger cloud.

(3) Stacking: When the small clouds join together, updrafts within the larger cloud increase. The updrafts near the center of the cloud are stronger than those near the edges. These updrafts cause the cloud body to grow vertically, so the cloud is stacked up. This vertical growth causes the cloud body to stretch into cooler regions of the atmosphere where drops of water and hail formulate and begin to grow larger and larger. When these drops of water and hail become too heavy for the updrafts to support them, they begin to fall from the cloud as rain, hail, etc.

Quranic Viewpoint

God said in the Quran:

-Have you not seen how Allah makes the clouds move gently, then joins them together then makes them into a stack, and then you see the rain come out of it...?

Chapter24: verse43

Meteorologists have only recently come to know these details of cloud formation, structure, and function by using advanced equipment like planes, satellites, computers, balloons, and the like to study winds and its direction, to measure humidity and its variations, and to determine the levels and variations of atmospheric pressure.

The preceding verse, after mentioning clouds and rain, speaks about hail and lightning:

-And He sends down hail from mountains (clouds) in the sky, and He strikes with it whomever He wills, and turns it from whoever He wills. The vivid flash of its lightning nearly blinds the sight.

Chapter24: verse43

Meteorologists have found that these cumulonimbus clouds, that shower hail, reach a height of 25,000 to 30,000 feet (4.7 to 5.7 miles), like mountains, as the Quran said: And He sends down hail from mountains (clouds) in the sky...

This verse may raise a question. Why does the verse say 'its lightning' in reference to the hail?

Does this mean that hail is the major factor in producing lightning? The book entitled "Meteorology Today", says that clouds become electrified as hail falls through a region in the cloud of supercooled droplets and ice crystals. As liquid droplets collide with hail, they freeze on contact and release latent heat. This keeps the surface of the hail warmer than that of the surrounding ice crystals.

When the hail comes in contact with an ice crystal, an important phenomenon occurs. Electrons flow from the colder object toward the warmer object. Hence, the hail becomes negatively charged.

The same effect occurs when super cooled droplets come in contact with a piece of hail and tiny splinters of positively charged ice break off. These lighter positively charged particles are then carried to the upper part of the cloud by updrafts. The hail, left with a negative charge, fall toward the bottom of the cloud, thus the lower part of the cloud becomes negatively charged. These negative charges are then discharged to the ground as lightning. We conclude from this that hail is the major factor in producing lightning.

This information on lightning was discovered only recently. Until 1600 AD, Aristotle's ideas on meteorology were dominant. For example, he said that the atmosphere contains two kinds of exhalation, moist and dry. He also said that thunder is the sound of the collision of the dry exhalation with the neighboring clouds, and lightning is the inflaming and burning of the dry exhalation with a thin and faint fire.

These are some of the ideas on meteorology that were dominant at the time of the Quran's revelation, fourteen centuries ago.

References

1. Holy Quran translated by Marmaduke Pickthall, George Allen and Unwin Ltd., London, Fifth Edition 1969.

Scientific Sources

- Http://www.islam-guide.com/.
- Meteorology Today, Ahrens, p. 437.
- The Works of Aristotle Translated into English: Meteorologica, vol. 3, Ross and others, pp. 369a-369b.
- Initial Cloud formation, by Carl Wozniak and the Great Lakes Collaborative, http://www.seaborg.nmu.edu/.

Medical Sciences in Holy Quran

- Embryology
- The Three Veils around Fetus
- Gender Determination
- Nature of Sperm
- The Sensory Characteristic of Skin
- Cerebrum
- Fingerprints
- Therapeutic Properties of Honey
- Menstruation

Embryology

وَلَقَدْ خَلَقْنَا الْإِنسَٰنَ مِن سُلَٰلَةٍ مِّن طِينٍ ۝ ثُمَّ جَعَلْنَٰهُ نُطْفَةً فِى قَرَارٍ مَّكِينٍ ۝ ثُمَّ خَلَقْنَا النُّطْفَةَ عَلَقَةً فَخَلَقْنَا الْعَلَقَةَ مُضْغَةً فَخَلَقْنَا الْمُضْغَةَ عِظَٰمًا فَكَسَوْنَا الْعِظَٰمَ لَحْمًا ثُمَّ أَنشَأْنَٰهُ خَلْقًا ءَاخَرَ ۚ فَتَبَارَكَ اللَّهُ أَحْسَنُ الْخَٰلِقِينَ ۝

يَٰٓأَيُّهَا النَّاسُ إِن كُنتُمْ فِى رَيْبٍ مِّنَ الْبَعْثِ فَإِنَّا خَلَقْنَٰكُم مِّن تُرَابٍ ثُمَّ مِن نُّطْفَةٍ ثُمَّ مِنْ عَلَقَةٍ ثُمَّ مِن مُّضْغَةٍ مُّخَلَّقَةٍ وَغَيْرِ مُخَلَّقَةٍ لِّنُبَيِّنَ لَكُمْ ۚ وَنُقِرُّ فِى الْأَرْحَامِ مَا نَشَآءُ إِلَىٰٓ أَجَلٍ مُّسَمًّى ثُمَّ نُخْرِجُكُمْ طِفْلًا ثُمَّ لِتَبْلُغُوٓا۟ أَشُدَّكُمْ ۖ وَمِنكُم مَّن يُتَوَفَّىٰ وَمِنكُم مَّن يُرَدُّ إِلَىٰٓ أَرْذَلِ الْعُمُرِ لِكَيْلَا يَعْلَمَ مِنۢ بَعْدِ عِلْمٍ شَيْئًا ۚ وَتَرَى الْأَرْضَ هَامِدَةً فَإِذَآ أَنزَلْنَا عَلَيْهَا الْمَآءَ اهْتَزَّتْ وَرَبَتْ وَأَنۢبَتَتْ مِن كُلِّ زَوْجٍۭ بَهِيجٍ ۝

The Creation of Man

In the Holy Quran the subject of human reproduction leads to a multitude of statements, which constitute a challenge to the embryologists, seeking a human explanation to them. It was only after the birth of the basic sciences, which were to contribute to our knowledge of biology, and especially after the invention of the microscope, that human was able to understand such statements. It was impossible for a man living in the early seventh century to have expressed such ideas. There is nothing to indicate that, at that time, men in the Middle East and Arabia knew anything more about this subject than the most enlightened European of this time.

Not one statement in Holy Quran contradicts today's data and, furthermore, none of the grossly mistaken concepts of human reproduction of the time has crept into the Quran:

-We created human of an extraction of clay[1];

-Then placed him as a drop (of seed) in a safe lodging;

-Then fashioned We the drop a clot, then fashioned We the clot a little lump, then fashioned We the little lump bones, then clothed the bones with flesh and then produced it another creation. So blessed be Allah, the Best of Creators!

Chapter23: verses12-14

-O mankind! If you are in doubt concerning the Resurrection, then lo! We have created you from a drop of seed, then from a clot, then from a little lump of flesh shapely and shapeless, that We may make it clear for you. And We cause what We will to remain in the wombs for an appointed time, and afterward We bring you forth as infants, then (give you growth) that ye attain your full strength. And among you there is he who dieth (young), and among you there is he who is brought back to the most abject time of life, so that, after knowledge, he knoweth naught[2].

Chapter22: verse5

In this section, we are going to revisit the verse 14 of chapter 23, and verse 5 of chapter 22 from the standpoint of embryology. Primarily, it is necessary to keep in mind that the following overview would be the author's presumption, which is based on today's science. The future scientific discoveries may declare some other facts in this regard, but nonetheless it is this author's contention that Holy Quran will indeed hold up to the test of time and future discoveries.

Scientific Viewpoint

Based on embryology, the period of pregnancy is divided into 3 main stages:

1-The first stage known as the zygotic stage starts at the time of fecundation. It lasts till the end of the third week. The clot in this stage called zygote consists of the morula and the Blastocyst.

2-The second stage known as the embryonic stage begins from the beginning of the fourth week. It endures tills the 8th week. In this stage, the zygote is known as the embryo.

3-The Third stage known as the fetal stage starts from the beginning of the third month. It lasts tills the time of birth. In this stage, the embryo formally becomes the fetus.

First to Third Week

The consecutive Division:

When the ovum fecundates a series of cell divisions occur in the womb. Consequently, a large number of cells come to existence with each new division. In the course of this process, these cells, called plastomer, are divided to smaller ones. 30 hours after fecundation, zygote has divided to become two cells; and 40 to 50 hours after fecundation it is now a 4-cell organism.

The zygote passes through the womb canal, in the interim of division. When it divides to be a 12-16 cellular organism (3rd or 4th day), it likens a berry seed and is called "morula".

Morula consists of a cellular group in the center, which is called the inner cellular accumulation; and a group in the circumference, called the outer cellular accumulation. Although, in this stage, all the morulas seem to be the same, discoveries have shown that the inner cellular accumulation shapes the real fetal tissues, and the outer cellular accumulation shapes the nutritional organism. As the interim, morula goes on dividing, a liquid from womb canal penetrates into the intercellular spaces.

Due to the continuous increase of this liquid, the intercellular spaces get enlarged and connect to each other. Ultimately, they form a unit hole, named as a blastocell.

In this stage, the veil over the zygote, which is called zunaplucida, is completely annihilated, and zygote is termed as blastocite. Henceforth, the inner cellular accumulation cells, called the embryoblast, adhere to each other and settle in one pole. In the interim, the outer cells of patrofoblasts become broad and form the blastocite wall.

It is from above process that the zygote passes the morula stage and the formation of blastosite till the end of the first week of completion and in the interim, completes the passage through the womb canal. It eventually embeds in the womb's mucus, striving to find its final resting-place for the term in the uterus.

According to the statements from the Quran previously mentioned, it seems that Quran entitles this period of pregnancy, as the "clot" since the uniform cells are suspended and adhered to the womb wall.

During the second and third of completion, blastosite penetrates the womb mucus and each of its two parts i.e. trophoblast and embryoblast are detached from each other by realigning their orientation. The trophoblast cells are deeply plunged into andometer and form the future offspring; and embryblast makes the three main layers of fetus, which are called the generative layers (Cambiums). These layers are:

1. Actoderm (external)
2. Mesoderm (middle)
3. Andoderm (internal)

Although, in this stage, the cambiums are formed, but no distinct human form is seen among the tissues. We presume that this stage is called as" lump of flesh" in verse 14 of chapter 23; because the fetus in this stage is similar to a piece of meat in which there is no distinct among its body members. Although, these three cambiums from which all the organs and tissues originate, have been formed.

Forth to eighth week: Embryonic Period

In this stage, the cambium starts changing and individuating; i.e. the formation of the specific tissues and organs begins and all of them become to take shape. The outer cambium (ectoderm) rises to form the brain, marrow, nerves, skin complexion with all its parts and the mucous membrane of mouth and nose.

Heart, blood, blood vessels, bones, muscles, kidneys, middle parts of skin and a part of the inner glands of middle layer (mesoderm). Also, the veil of digestive and respiratory system, thyroid gland, liver, pancreas, are generated by endoderm.

How does really this happen? How do uniform blastocite cells generate all these amazing and sophisticated systems, different tissues and the thinking organs? This scientific procedure of fetus' creation really astonishes the world's great scientists and teaches us that God almighty is indeed the unique creator of the universe!

The Holy Quran amazingly points to this subject in verse 14 of chapter 23, to have us think about His majesty where it says:

-Then fashioned We the little lump bones, then clothed the bones with flesh and then produced it another creation, so blessed be Allah, the best of creators!

It seems, in verse 5 of chapter 22, this stage is named as " lump of flesh shapely", since, in this stage, systems and body tissues are produced and the fetus is really shaped.

From Third to Tenth Month: Fetal Period

From the beginning of the third month, the body growth is rapid, while the distinction between the tissues and internal organs are less significant. The heartbeat is one of the most important events occurring after the completion of the third month. At this time, the fetus' motion inside the womb begins because of the formation of nerves. The pregnant mother feels her fetus' motion at this time[3].

With no further explanation required, we draw your attention to these glorious verses:

-What ails you that you look not for Majesty in God,
-Seeing He created you by stages!

Chapter 71: verses 13 and 14

Quranic Viewpoint

A few years ago, a group of men in Riyadh, Saudi Arabia collected all of the verses in the Quran, which discuss embryology, the growth of the human being in the womb. They said, "Here is what the Quran says. Is it the truth?" In essence, they took the advice of the Quran: "Ask those who know." They were able to approach, by chance, a non-Muslim who is a professor of embryology at the University of Toronto, Keith Moore, Ph.D. and author of textbooks on embryology as well as a world expert on the subject. They invited him to Riyadh and said, "This is what the Quran says about your subject. Is it true? What can you tell us?"

While staying in Riyadh, they gave Dr. Moore extensive help in translation as well as all of the cooperation for which he asked. Dr. Moore was extremely surprised at what he found that he changed his textbooks. In fact, in the second edition of one of his books, 'Before We Are Born', he included some material in the section about the history of embryology. This information he included was based on what he found in the Quran shedding light on the subject. This information was ahead of its time and had not been available outside those who knew the content of the Holy Quran.

When Dr. Keith Moore was interviewed for a television presentation, he mentioned the information the Quran states about the growth of the human being, which was not known until thirty years ago. In fact, he said that one item in particular -the Quran's description of the human being as a "leech-like clot" (*'alaqah*) at one stage (Chapter22: verse5; chapter23: verse14; and chapter40: verse67). This information had never occurred to him in this manner. It was new to him; but when he checked on it, he found that it was true, and so he added it to his book. He said, "I never thought of that before" and he went to the zoology department and asked for a picture of a leech. When he found that it looked just like the human embryo, he decided to include both pictures in one of his textbooks.

Although, the aforementioned example of man researching information contained in the Quran deals with a non-Muslim, it is still valid because he is one of those who is knowledgeable in the subject of embryology. Had a man of lesser caliber claimed that what the Quran says about embryology is true, then one would not necessarily have to accept his word. However, because of the high position, respect, and esteem man gives scholars, one naturally assumes that if they research a subject and arrive at a conclusion based on that research, then the conclusion is valid.

Skeptic's Reaction

Dr. Moore also wrote a book on clinical embryology. When this information was included in a presentation he made in Toronto, it caused quite a stir throughout Canada. It was on the front pages of some of the newspapers across Canada, and some of the headlines were quite funny. For instance, one headline read: "SURPRISING THING FOUND IN ANCIENT PRAYER BOOK!" It seems obvious from this example that people indeed were impressed by the remarkable discovery. In fact, one newspaper reporter asked Professor Moore, "Don't you think that maybe the Arabs might have known about these things -the description of the embryo, its appearance and how it changes and grows (long before we did)? Maybe they were not scientists, did (they perform) some crude dissections on their own?"

When the professor had given the presentation he had included some slides of the particular stages of development that the Quranic statement related to. Upon hearing this question, the professor immediately pointed out to him that he [i.e., the reporter] had missed a very important point - all of the slides of the embryo that had been shown and that had been projected in the film had come from pictures taken through a microscope. He said, "It does not matter if someone had tried to discover embryology fourteen centuries ago. They could not have seen it!"

All of the descriptions in the Quran regarding the embryo appearance are when the embryo is still too small to see with the eye. One needs a microscope to examine them at this early stage. Since such a device had only been around for little more than two hundred years earlier. Dr. Moore taunted, "Maybe fourteen centuries ago someone secretly had a microscope and did this research, making no mistakes anywhere. Then he somehow taught Muhammed (as) and convinced him to put this information in his book. Then he destroyed his equipment and kept it a secret forever. Do you believe that? You really should not unless you bring some proof because it is such a ridiculous theory."

In fact, when he was asked, "How do you explain this information in the Quran?" Dr. Moore's reply was, "It could only have been divinely revealed!"

A Scientist's Interpretation of Embryology in the Quran

We offer you the following interpretation made by Dr. Keith L. Moore, Ph.D., F.I.A.C., professor of anatomy and associate dean basic sciences, faculty of medicine, university of Toronto, Ontario M55 IAB, Canada[4]:

"Statements referring to human reproduction and development are scattered throughout the Quran. It is only recently that the scientific meaning of some of these verses to be fully appreciated. The long delay in interpreting these verses correctly resulted mainly from a lack of awareness of scientific knowledge combined with from inaccurate translations and commentaries.

Interest in explanations of the verses of the Quran is not new. People used to ask the prophet Muhammed (May peace be upon him) all sorts of questions about the meaning of verses referring to human reproduction. The Apostle's answers form the basis of the Hadith literature:

*-Then We placed him as a drop in a place of rest.**

This statement is from Chapter23, verse13. The drop or nutfah has been interpreted as the sperm or spermatozoon, but a more meaningful interpretation would be the zygote, which divides to form a blastocyst, which is implanted in the uterus ("a place of rest"). This interpretation is supported by another verse in the Quran, which states that "a human being is created from a mixed drop." The zygote forms by the union of a mixture of the sperm and the ovum ("The mixed drop").

-Then of that leech-like structure, We made a chewed lump.

This statement is from Chapter23, verses14. The Arabic word "mudghah" means "chewed substance or chewed lump." In the early part of the fourth week, the embryo is just visible to the unaided eye because it is smaller than a kernel of wheat. Toward the end of the fourth week, the human embryo looks somewhat like a chewed lump of flesh. The chewed appearance results from the somites, which resemble teeth marks. The somites represent the beginnings or primordia of the vertebrae.

-Then We made out of the chewed lump, bones, and clothed the bones in flesh.

This continuation of the verse14 0f chapter23 indicates that out of the chewed lump stage, bones and muscles form. This is in accordance with embryological development. First the bones form as cartilage models and then the muscles (flesh) develop around them from the somatic mesoderm.

-Then We developed out of it another creature.

This next part of the verse implies that the bones and muscles result in the formation of another creature. This may refer to the human-like embryo that forms by the end of the eighth week. At this stage it has distinctive human characteristics and possesses the primordia of all the internal and external organs and parts. After the eighth week, the human embryo is called a fetus. This may be the new creature to which the verse refers.

-And He gave you hearing and sight and feeling and understanding.

This part of verse9 of chapter32 indicates that the special senses of hearing, seeing, and feeling develop in this order, which is true. The primordia of the internal ears appear before the beginning of the eyes, and the brain (the site of understanding) differentiates last.

-Then out of a piece of chewed flesh, partly formed and partly unformed...

This part of verse5 of chapter22 seems to indicate that the embryo is composed of both differentiated and undifferentiated tissues. For example, when the cartilage bones are differentiated, the embryonic connective tissue or mesenchyme around them is undifferentiated. It later differentiates into the muscles and ligaments attached to the bones.

-And We cause whom We will to rest in the wombs for an appointed term.

This next part of verse5 of chapter22 seems to imply that God determines which embryos will remain in the uterus until full term. Many embryos abort during the first month of development, and only about 30% of zygotes that form, develop into fetuses that survive until birth. This verse is also interpreted means that God determines whether the embryo will develop into a boy or girl.

The interpretation of the verses in the Quran referring to human development would not have been possible in the 700 AD, or even a hundred years ago. We can interpret them now because the science of modern Embryology affords us new understanding. Undoubtedly there are other verses in the Quran related to human development that will be understood in the future as our knowledge increases."

The Embryonic Phases

Dr. G.C. Goeringer is course director and associate professor of medical embryology at the department of cell biology, school of medicine, Georgetown University, Washington. D.C. Sheik Abdul Majid Zindani, a professor of Islamic studies in King Abdulaziz University in Jeddah, Saudi Arabia, says about his meeting with Dr. Goeringer[5]:

"We met him and asked him whether in the history of embryology there was any mention of the different stages of embryonic development and whether there were any books on embryology at the time of the Prophet Muhammed (Peace Be upon him) or the centuries after him which mention these various stages, or whether the division into these different stages only came to be known in the middle of the nineteenth century. He said that the ancient Greeks were concerned with the study of embryology and many of them attempted to describe what happens to the fetus and how it develops. We agreed with him that Aristotle, among others, attempted to expound some theories on the subject, but was there any mention made of these stages?

We know that these stages were not known until the middle of the nineteenth century and were not proven until the beginning of the twentieth century. After a long discussion, Professor Goeringer concurred that there was no mention of these phases. Thus, we asked him if there was any specific terminology applied to these phases similar to that found in the Quran. His reply was negative. We asked him: 'What is your opinion on these terms, which the Quran uses to describe the phases, which the fetus goes through?'

After long discussions, he presented a study at the 8th Saudi Medical Conference. He mentioned in the study man's basic ignorance of these phases. He also discussed the comprehensiveness and precision of these Quranic terms in describing the development of the fetus by means of concise and comprehensive terms, which convey far-reaching truth. Let us listen to Professor Goeringer explains his opinion:

"In a relatively few aayahs (Quranic verses), is contained a rather comprehensive description of human development from the time of the commingling of the gametes through organogenesis. No such distinct and complete record of human development such as classification, terminology, and description existed previously. In most, if not all, instances, this description antedates by many centuries the recording of the various stages of human embryonic and fetal development recorded in the traditional scientific literature."

The discussion with Professor Goeringer leads us to talk about a fact which discovered recently and which will eliminate any controversy. Although the virgin birth of the prophet Jesus Christ (peace be upon him) has been a Christian belief for centuries, some among the Christians insist that Christ must have had a father, as a virgin birth is "scientifically impossible". They argue this, and perhaps they do not know that there could be a creation without a father. The Quran replied to them and used as an example the creation of Adam. Allah said:

-The similitude of Jesus before Allah is as that of Adam; He created him from dust, then said to him: "Be": And he was.

Chapter3: verse59

There are three types of creation:
1. Adam, who was created without a mother or father.
2. Eve, who was created without a mother.
3. Jesus Christ, who was created without a father.

Therefore, the One who was able to create Adam without a father or a mother is also able to create Jesus from a mother and without a father. In spite of this, Christians continue to argue even though Allah has sent them evidence after evidence and proof after proof. When they are asked why they persist in this contention, they reply that they have never seen or heard of anybody being created without a father and a mother.

Now modern science revealed that many animals and beings in this world are born and reproduced without fertilization from the male of the species. For example, a male bee is no more than an egg, which is not fertilized by the male, whereas the egg, which has been fertilized by the male, functions as a female. Moreover, male bees are created from the eggs of the queen but without fertilization by a male.

There are many other examples such as this in the animal world. Moreover, man today has the scientific means of stimulating the female's egg of some organisms so that this egg develops without fertilization by a male.

Let us read the words of Professor Goeringer:

"In another type of approach, unfertilized eggs of many species of amphibians and lower mammals can be activated by mechanical (such as pricking with a needle), physical (such as thermal shock) or chemical means by any of a number of different chemical substances, and continue to advance to stages of development. In some species, this type of parthenogenetic development is natural."

Allah has given us the definitive answer and he used Adam, whom they believe in, as an example of a human being who has no father or mother. The Christians regard as deviance the fact that a human being can be born without a father. Thus, Allah has shown them an analogy of a human being who had no father and no mother, that is, Adam. The Quran says:

-The similitude of Jesus before Allah is as that of Adam; He created him from dust, then said to him: "Be": and he was.
Chapter3: verse59

Allah has willed that there be such scientific advancements and discoveries which provide proof after proof of the truth which has been revealed in the Quran. It is in this way that the verses of this glorious book were revealed with the passage of time. The verses are known to the foremost scholars and scientists of our religion and of generations to come. Science will never deplete the wonders of the Quran.

-And those to whom knowledge has come see that the (Revelation) sent down to you from your Lord - that is the truth, and that it guides to the path of the Exalted (In Mighty), worthy of all praise.
Chapter34: verse6

Allah, may He be Exalted and Glorified, said in the Quran:

-And you shall certainly know the truth of it (all) after a while.
Chapter38: verse88

Allah also said:
-For every prophecy is a limit of time, and soon shall you know it.
Chapter6: verse67

And He said:

-Soon will we show them our signs in the (furthest) regions (of the earth), and in their own souls, until it becomes manifest to them that this is the truth. Is it not enough that your Lord does witness all things?
Chapter41: verse53

(*) Please note: The translations of the verses of the Quran in the above paper were provided by Sheik Abdul Majid Zendani, a professor of Islamic studies in King Abdulaziz University in Jeddah, Saudi Arabia, the links to the translations in this page are from Yusuf Ali.

References
1. Holy Quran translated by Arthur J. Arberry, Ansarian Publishers, Qum, the Islamic Republic of Iran, 1999.
2. Holy Quran translated by Marmaduke Pickthall, George Allen and Unwin Ltd., London, Fifth Edition, 1969.
3. Quran and the New Medicine (in Persian language), By Doctor M. Noori Shadkaam, Dar Al-kotob Al-Islamiah Publishers, Teheran, 1999.
4. Keith L. Moore, Ph.D., F.I.A.C., the Department of Anatomy, University of Toronto, Canada, from the Journal of the Islamic Medical Association, Vol.18, Jan-June 1986, pp.15-16.
5. Embryonic Phases, http://www.it-is-truth.org/contents.htm.

The Three Veils around Fetus

خَلَقَكُم مِّن نَّفْسٍ وَاحِدَةٍ ثُمَّ جَعَلَ مِنْهَا زَوْجَهَا وَأَنزَلَ لَكُم مِّنَ الْأَنْعَامِ ثَمَانِيَةَ أَزْوَاجٍ يَخْلُقُكُمْ فِي بُطُونِ أُمَّهَاتِكُمْ خَلْقًا مِّن بَعْدِ خَلْقٍ فِي ظُلُمَاتٍ ثَلَاثٍ ذَٰلِكُمُ اللَّهُ رَبُّكُمْ لَهُ الْمُلْكُ لَا إِلَٰهَ إِلَّا هُوَ فَأَنَّىٰ تُصْرَفُونَ ۝

مَّا لَكُمْ لَا تَرْجُونَ لِلَّهِ وَقَارًا ۝ وَقَدْ خَلَقَكُمْ أَطْوَارًا ۝

The following verses of the Holy Quran inspired a Canadian scientist to convert to Islam after contemplating them:

-...He makes you in the wombs of your mothers in stages one after another in three veils of darkness. Such is Allah your Lord and cherisher; to Him belongs (all) domination. There is no God but He; then how are ye turned away (from your true center)[1]?

Chapter 39: verse 6

-What ails you, that you look not for majesty in God, seeing He created you by stages[2]?!

Chapter 71: verse 13 and 14

Let us glance over his opinion about the context of this verse, as an established and renowned embryologist.

-...He makes you in the wombs of your mothers in stages one after another in three veils of darkness...

This statement is from Chapter 39, verse 6 from the Holy Quran. We do not know when human beings first realized that development occurred in the uterus (womb). Leonardo da Vinci drew the first known illustration of a fetus in the uterus in the 15th century. In the 2nd century AD, Galen described the placenta and fetal membranes in his book "On the Formation of the Fetus." Consequently, doctors in the 7th century AD, likely knew that the human embryo developed in the uterus. It is unlikely that they knew that it developed in stages, even though *Aristotle* had described the stages of development of the chick embryo in the 4th century BC. The realization that the human embryo develops in stages was not discussed and illustrated until the 15th century.

After the microscope was discovered in the 17th century by *Leeuwenhoek* descriptions were made of the early stages of the chick embryo. The staging of human embryos was not described until the 20th century. *Streeter* (1941) developed the first system of staging, which has now been replaced, by a more accurate system proposed by *O'Rahilly* (1972).

Common sense dictates that when you envelope an object by placing it inside a containment, it becomes relatively dark inside the containment, then if this contained object is placed further inside two more layers it will become even darker inside.

The three-fold darkness may refer to:

i) Anterior abdominal wall
ii) The uterine wall
iii) Amnio Chorionic membrane

Although there are other interpretations of this statement, the one presented here seems the most logical from an embryological point of view[3].

To achieve a better understanding of the matter and its correlation with the corresponding verse, you consider the following scientific passage.

Scientific Viewpoint

From the standpoint of embryology, three layers envelop the fetus which is as follow respectively from outer to inner layer[4]:

1) *The womb rims (fetal "membrane" membranes)*: The fetus is placed inside it; and its function is to provide blood for the fetus and consequently to feed and to excrete the waste.

These membranes protect the embryo and provide for its nutrition, respiration, and excretion: the yolk sac (umbilical vesicle), allantois, amnion, chorion, decidua, and placenta. *(Miller-Keane Medical Dictionary, 2000)*

2) *Chorion Layer:* A membrane surrounding the fetus (embryo) within the womb (uterus).

3) *Amnion Layer:* This layer surrounds the Amniotic cavity

(The closed sac between the embryo and the amnion, containing the amniotic fluid), which is full of liquid (Amnion).[5]

The albuminous fluid contained in the amniotic sac; called liquor amnii or more informally, waters. The fetus floats in the amniotic fluid, which serves as a cushion against injury from sudden blows or movements and helps maintain a constant body temperature for the fetus. Normally the fluid is clear and slightly alkaline; discoloration or excessive cloudiness may indicate fetal distress or disease, as in erythroblastosis fetalis in which fluid is usually greenish yellow. The amount varies from 500 to 1500 ml.

An excessive amount of amniotic fluid (more than 2000 ml) is called hydramnios; the amount may be as much as several gallons. The cause of this condition is unknown but it frequently accompanies multiple pregnancy or some congenital defect of the fetus, especially hydrocephalus and meningocele.

Oligohydramnios in which there may be less than 100 ml of fluid present refers to an abnormally small amount of amniotic fluid. The cause is also unknown. The condition may produce pressure deformities of the fetus, such as clubfoot or torticollis. Adhesions may result from direct contact of the fetus with the amnion.

Removal of a sample of amniotic fluid from the pregnant uterus is called "amniocentesis".

Smooth chorion is formed by gradual degeneration and disappearance of the chorionic villi. Villus chorion maintains its chorionic villi and ultimately gives rise to placenta. Placenta is a lobulated structure which contains cotyledons. Blood circulates through the intervillous spaces of the cotyledons.

Quranic Viewpoint

We present to you Dr. Marshall Johnson, Professor Emeritus of Anatomy and Developmental Biology at Thomas Jefferson University Philadelphia,

Pennsylvania, USA. For 22 years, he was Professor of Anatomy, the Chairman of the Department of Anatomy, and the Director of the Daniel Baugh Institute. He was also the President of the Teratology Society. He has authored more than 200 publications.

Sheik Abdul Majid Zindani, a Professor of Islamic Studies in King Abdulaziz University in Jeddah, Saudi Arabia discusses his meeting with the professor Johnson[6]:

"We first met with Professor Johnson at the 7th Saudi Medical Conference, where a special committee was formed to investigate scientific signs in the Quran and the Sunnah.

When we met together with the committee, Professor Johnson asked us what our committee was doing. We told him that the subject of our study was the relationship between what the Quran and Sunnah have contained 1400 years ago and what modern scientists tell us. He asked: Like what? We said: 'For example, modern science tells us that the human prenatal development goes through several stages whereas the Quran mentioned these stages to us 1400 years ago.'

Professor Johnson was appalled when he heard this. He expressed his feeling by simply saying: 'No' no, no! What kind of talk is that?'

The readers perhaps can understand the effect of such statements had on him. Professor Johnson is one of the foremost scientists in the United States. He knew, of course, that after the discovery of the microscope in the 16th century, physicians throughout the 17th century believed that a human being's origin began entirely in the male's semen, specifically in the male's sperm.

What was first observed by the microscope was the evidence used by scientists in the 17th century, and partly in the 18th century, to support their belief that the human being was created wholly from the male sperm, but after the discovery of the ovum, they found it larger than the sperm of the male. In this way they ignored the role of the man in the 18th century, just as they had ignored that of the woman in the 17th century.

It was not until the middle of the 19th century when the scientists began to discover that the human embryonic development took place in several successive stages.

The Sheik continues: "It was for this reason when Professor Johnson was told that this information has been in the Quran for 1400 years that he stood up and shouted: *'No, No!'* So we in turn handed him a copy of the Quran and showed him the following verse, which he read in the English translation:

-What is the matter with you, that you are not conscious of Allah's Majesty. Seeing that it is He who has created you in diverse stages?

Chapter71: verse13 and 14

Then we showed him this verse:

-He creates you in the wombs of your mother, in stages, one after another, in three veils of darkness.

Chapter 39: verse 6

At this, Professor Johnson sat down and said: But this could be explained in three possibilities. The first is that it can be mere coincidence.

...We collected more than 25 texts and presented them to him. Then we asked him: 'Is it possible that these texts are coincidence?' Moreover, the Glorious Quran has given each of these stages a name: the first being the *nutfah*, or the drop of water, the second being the '*alaqah*, a leech-like substance, the third being the *mudghah*, or a chewed-like lump, the fourth being bones, and then the clothing of the bones with flesh. Can all this be a coincidence? He flatly said: 'No.'

Then we asked him: 'Then what remains?' He said: The other possibility is that Muhammed had hold [of a] powerful microscope.

Professor Johnson was challenged: 'you know that very powerful microscopes can only obtain this sort of minute and specialized knowledge, which has been contained in the Quran. And anyone possessing such powerful microscopes must also possess very high technology, which must be reflected in his daily living, his house, his food, his control and management of war and quest for peace...etc. And you know that technological advancement is a process of cumulative inheritance, passed in and proved upon from one generation to the other.'

As the Sheik reports: "Professor Johnson laughed and said: In fact I saw the first microscope invented in the world. It does not magnify more than 10 times and does not even show a clear picture."

AS the reader knows, the Prophet Muhammed (Peace Be upon Him), never had possession of scientific equipment or microscopes. The only thing that remains for us to say that he was a Messenger from Allah.

After this, Professor Johnson began to take interest in the study of scientific signs in the Quran, concentrating in some of his research on the stages of embryonic development. While Dr. Moore and others discussed the external fetal appearance, Professor Johnson concentrated his presentation on the Quran's detailed descriptions of the internal as well as external of the fetus.

As Professor Johnson stated: "In summary, the Quran describes not only the development of external form, but emphasizes also the internal stages, the stages inside the embryo, of its creation and development, emphasizing major events recognized by contemporary sciences."

How then would we describe this embryo? What do we say? Could we say it is the complete creation? Then we are describing the part which is already created, and if we say it is an incomplete creation then we are describing the part which is not yet created, the question would be: Is it a complete creation or is it an incomplete creation? There is no better description of that stage of embryologenesis than the Quranic description, which says:

-...mudghah (chewed-like structure) partly formed and partly unformed...

Chapter22: verse5

As Professor Marshall Johnson states in giving the conclusion of his research: "As a scientist, I can only deal with things, which I can specifically see. I can understand the words that are translated to me from the Quran. As I gave the example before, if I were to transpose myself into that era, knowing what I knew today in describing things, I could not describe the things, which were described. I see no evidence for the refutation of the concept that this individual, Muhammed, had to be developing this information from some place."

He continues on: "So I see nothing here in conflict with the concept that divine intervention was involved in what he was able to write."

Yes, it is the most supreme revelation. The only way left to mankind is to follow the example of those great scientists, acknowledging that Allah has revealed to Muhammed (Peace Be upon Him), a Book from Allah's knowledge. Allah has further promised that mankind will, over time, come to discover the signs which prove that the Quran is truly a Book revealed by Allah.

In what other book than Quran indeed we could find such an exact indication of fetus' triple veils in the womb, while there was no evidence for fetus genesis 1400 years ago? Is not this strong proof for divinity of Holy Quran for those who are in doubt? Those who possess strong minds and pure logic must surely submit to the uncontroversial proof.

Terminology

Uterus: (or womb) A hollow muscular organ into which the ovum is received through the fallopian tubes, and where it is retained during development, and from which the fetus is expelled through the cervical os (uterine neck) into the vaginal canal.

Fetus: The developing human from three months after conception to birth.

Embryo: The developing human individual from the time of implantation to the end of the eighth week after conception.

Amniotic fluid: The fluid within the amnion that bathes the developing fetus and protects it from mechanical injury. (Miller-Keane Medical Dictionary, 2000)

(am´ne-on): the innermost fetal membrane, which forms a fluid-filled sac that surrounds the embryo and later the fetus; as it enlarges it gradually obliterates the chorionic cavity and enfolds the umbilical cord. (Miller-Keane Medical Dictionary, 2000)

References
1. Holy Quran translated by Marmaduke Pickthall, George Allen and Unwin Ltd., London, Fifth Edition 1969.
2. Holy Quran translated by Arthur J. Arberry, Ansarian Publication, Qum, the Islamic Republic of Iran, 1999.
3. Keith L. Moore, Ph.D., F.I.A.C., the Department of Anatomy, University of Toronto, Canada.
4. Quran and the New Medicine (In Persian), by Dr. M. Noori Shadkaam, Dar Al-Kotob Al-Islamiah, Qum, the Islamic Republic of Iran, 1999.
5. Miller-Keene Medical Dictionary, 2000.
6. http://www.islam-guide.com/.

Gender Determination

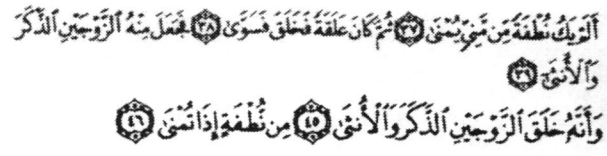

In this day and age, it is common fact that sperms are the deciding factors in determining the gender (male or female) in a new embryo. This determination through the male sperm is due to the fact that sperms have an X and Y-chromosomes, while the female ovum has only X-chromosomes (X, X).

Does the Quran state anything in this regard? Interestingly enough, the concept of "Gender Determination" as being described by geneticists, is one of the outstanding information revealed to us in the Quran. Through the Quran, some verses inform us that it is the male's sperm, which determines the type of gender for the new offspring:

-*Was he not a sperm-drop spilled?*
-*Then he was a blood clot and He created and formed.*
-*And He made of him two kinds, male and female¹.*
Chapter75: verse37-39

-*[And] He has created both sexes, male and female,*
-*From a drop of semen, which has been ejected [ejaculated]².*
Chapter53: verse45 and 46

In order to perceive the elegance of these Quranic statements, you may take the following scientific passage into deep consideration.

Scientific Viewpoint

Normally only one ovum is monthly spawned from women's ovary, whereas, millions of sperms exist inside men's semen, which are released to fertilize the ovum. When one of these sperms first reaches the ovule, it penetrates the cell membrane and both nucleiis come together to form a complete and fertilized cell. This cell divides into two other cells; and each of these two cells are also divided into two others; and this repetitive procedure continues until there are millions of cells. This initial division though is the first step in fetus genesis.

A man's semen consists of male and female sperm in an equal part. If a sperm containing female chromosomes first reaches the ovary, the offspring will be female; and if the sperm carries information for a male, then the offspring will be male.

To gain a better understanding, it is first necessary to explain the formation of male and female sexual cells. Then we will state the fecundity of these cells and the formation of the male and female fetus.

Formation of Male and Female Sex Cells

The human body consists of billions of cells. Each of these cells comprises 46 chromosomes or 23 pairs, which 22 pairs of them are identical and called autosomes. The remaining pair contains the sex chromosomes whose combination for female sex chromosome is XX and for male sex chromosome is XY.

These pairs of chromosomes are constant among the human cells and are the agent of distinction between human and the other living creatures. One of the two members of each pair of chromosome is passed down by inheritance from father and the other one from mother. For survival, the number of sex chromosomes is imperative to be half of the number of somatic chromosome cells, i.e. 46 chromosomes must be decreased to 23, otherwise, it would cause to bear a child with twice as the number of chromosomes as his or her parents. This decrease termed "meiosis division" which occurs in both women's ovary and men's testis. Because of this division, the entire set of chromosomes produced in the ovary are identical and carry form only X-chromosomes to assist in the determination of gender.

However, half of the chromosomes produced in testis are x type and the other half are y type. The abstract of this division is as follows:

44 + xx {22 + x, 22 + x} Ovary
44 + xy {22 + x, 22 + y} Testis

Fecundity

Fecundity is the combination of female's ovum (22 + x) and male's sperm (22 + x) or (22 + y). Since, half of the sperms contain x sex chromosomes and the other half contains y type sex chromosomes, the fetus will be female, when a sperm enters the ovum carrying an x sex chromosome. Should the ovum be fertilized by a Y carrying sperm it will be male.
That is:

(22 + x) + (22 + x) = 44 + xx female
(22 + x) + (22 + y) = 44 + xy male[3]

Quranic Viewpoint

By paying attention to the above scientific viewpoint, we come to this conclusion that it is the sperm, which determines the gender of fetus, because it contains both x and y-chromosomes; and this is exactly what Quran stated 14 centuries ago.
We can now well perceive the intention of these Quranic statements:

-[And] He has created both sexes, male and female,
-From a drop of semen, which has been ejected [ejaculated].
Chapter53: verse45 and 46

The male sperm is the "ejaculated drop". Females do not possess this ability and it lies solely with the male sperm. The male sperm carrying either an X or Y chromosome determines sex. Therefore, the ejaculated drop determines sex[4].
Unfortunately, because of some deviated traditional habits existing in some communities, many men divorce their wives for bearing female children! This is because they illusively think that women are responsible in gender determination. Reflection on the actual scientific evidence indicates that it is their own sperm, which determines the gender of their children, not their wives' ovum. Women are as a fertile tilth, which raise what is cultivated in. This is exactly what has been revealed in verse 223

of chapter 2:

-Your wives are a tilth for you (to cultivate)...

Thus the Quran's glorious wisdom is manifested here:

-and He made of him two kinds, male and female.

The pronoun "him" refers to the sperm, which the gender of human hinges upon. These verses state that the offspring's generic origin is man's sperm, regardless of being male or female and this is a scientific fact, which has been inspirationally mentioned in Holy Quran.

Fourteen hundred years ago there were no microscopes or other scientific equipment by which Muhammed (Peace Be Upon Him) could perceive this information. The nature of the sperms and chromosomes could not be evident to him on any level. How did he come to reveal these wonderful statements about the gender determination? We have to bear in mind that he lived among the most backward nations who used to bury their daughters alive. The only logical explanation, which remains is that Holy Quran could have not logically had a human source; and that Allah, the Almighty Himself is undoubtedly, the Author of Quran.

References
1. Holy Quran translated by Arthur J. Arberry, Ansarian Publishers, Qom, the Islamic Republic of Iran, 1999.
2. Holy Quran translated by Thomas Ballantine Irving, Suhrawardi Research and Publication Center, Teheran, 1998.
3. Quran and the New Medicine (in Persian language), By Doctor M. Noori Shadkaam, Dar Al-kotob Al-Islamiah, Teheran, 1999.
4. Quran, a Teacher to Modern Scientists, by Sabeel Ahmed, Co-chairman of the Da'wa Committee and Board of Director at the Muslim Community Center, Illinois, July 1997, http://www.iol.ie/.

Nature of Sperm

أَلَمْ يَكُ نُطْفَةً مِّن مَّنِيٍّ يُمْنَىٰ ۝ ثُمَّ كَانَ عَلَقَةً فَخَلَقَ فَسَوَّىٰ ۝

Science has discovered that man's semen is full of sperm, which are not visible with the naked eyes, but are observable by a microscope. Each sperm is a single cell, consisting of a head, neck and a tail, which resembles a leech. Of course, this knowledge came to us after the invention of microscope in 17th century by Robert Hooke (1635-1703).

Some verses of the Holy Quran illustrate the stages of human embryonic development in the womb, since the time of fecundity till the birth of child. However, among these Quranic explanations about the fetus genesis and its development, (See "Embryology" and "The Three Veils around Fetus") the description of sperm and its individual entity is quite brilliant. We may take the following verses into full consideration:

-Was he not a drop of sperm emitted?
-Then did he become a leech-like clot...[1]?

Chapter75: verse37 and 38

The Holy Quran vividly states that all human geneses originated from a leech-like clot. This verse speaks of the existence of sperm in the drop of semen and even describes its shape as being like a leech. At the first glance, one is convinced that this unique simile clarifies the majesty of the Quran as a miracle, especially after describing without the benefit of modern science and technology, that sperm is indeed similar to leech, in both appearance and function. We have to keep in mind that this statement was made 1000 years preceding the invention of microscope.

Scientific Viewpoint

After the discovery of the microscope in the 16th century, physicians throughout the 17th century believed that a human being's origin began entirely in the male's semen, specifically in the male's sperm. What they saw by using the microscope, was the evidence used by scientists in the 17th century (and partly in the early 18th century), to support their belief that the human being was created wholly from the male sperm. After the discovery of the ovum, which they found to be larger than the sperm of the male they reversed this theory. In this way, they ignored the role of man in the later 18th century, just as they had ignored that of woman in the 17th century.

It was not until the middle of the 19th century when the scientists began to discover that the human embryonic development took place in several successive stages (See "The Three veils around fetus).

Quranic Viewpoint

In embryology, the description that Holy Quran uses seems to fit quite well with what we know now about the early developments of the embryo -"leech-like thing", "clinging thing", "chewed-like thing", etc. The word 'alaqah has three meanings in Arabic[2]:
A) leech
B) a suspended thing
C) a blood clot

In comparing the fresh-water leech to the embryo at the 'alaqah' stage, a great similarity has been found. We can understand the second meaning in the attachment of the embryo to the womb of the mother, in the same way a leech clings to the skin. Furthermore, during the 'alaqah' stage (days 7-24), the blood is caught within blood vessels and that is why the embryo acquires the appearance of a blood clot, in addition to its leech-like appearance[3]. Since, just as a leech derives blood from its host, the embryo derives blood from the womb. The embryo of 23-24 days in its appearance remarkably resembles a leech. As there were no microscopes available 1400 years back, such a statement in the Quran is fascinating and miraculous indeed[4].

In the interim, this clinging blood clot suspends itself from the wall of the womb securing needed nutrients and sustenance from it. All three descriptions of sperm are given by a single Quranic word: 'alaqah'.

It is obvious that this simile discovers the truth and comes from beyond the confines of the existing human's wisdom of the time. This verse is a conspicuous sign of miraculous aspects of Holy Quran. Science discovered this fact after the invention of microscope by Robert Hooke, namely, hundreds of years after the revelation of this verse. This invention disclosed how a human being is brought into existence from these leech-like sperms.

However, the presence of such knowledge in Holy Quran may raise a question about the source of such an amazing statement. An unlettered Muhammed (Peace Be upon Him), whose first job had been watching the sheep in the desert of Arabian Peninsula could have not logically known that a human is originated from a tiny sperm, which can not be seen by naked eyes. He also would not have the resources to recognize that the sperm is indeed similar to a leech. He also would not have known that the fertilized ovum functions like a clinging substance in the womb. With having no scientific equipment or microscope, these discoveries were beyond the scope of any man at that time in history without heavenly intervention. The only thing that remains to be deduced by these facts is that Holy Quran is a perpetual divine miracle and Muhammed (Peace Be Upon Him) was a Messenger from God.

References

1. Holy Quran translated by Abdullah Yusuf Ali, Dar Al Arabia Publication, Beirut, Lebanon, 1968.
2. Encyclopedia of Quran (Qaamousse Quran), by Ali Akbar Qurashi, Dar Alkutub Al-Islamiah/ See Also: Monjid Al-Tollab (An Arabic-Arabic Encyclopedia), by Fu'aad Afraum Al-Bostaani.
3. Science and the Quran, by Noori Al-Ali, Youth Today Newsletter, Second Edition, London, 2001, http://www.islamsays.org.uk/.
4. Adapted from Dr. Keith L. Moore, Professor of Embryology, Department of Anatomy, Univ. of Toronto, Canada, recipient of numerous awards and honors, including in 1984, the J.C.B. grand award, which is the highest honor, granted by the Canadian Association of Anatomists.

The Sensory Characteristic of Skin

إِنَّ ٱلَّذِينَ كَفَرُواْ بِـَٔايَٰتِنَا سَوْفَ نُصْلِيهِمْ نَارًا كُلَّمَا نَضِجَتْ جُلُودُهُم بَدَّلْنَٰهُمْ جُلُودًا غَيْرَهَا لِيَذُوقُواْ ٱلْعَذَابَ إِنَّ ٱللَّهَ كَانَ عَزِيزًا حَكِيمًا ۝

By paying a considerable amount of attention to the following verse, one can apprehend the existence of a wonderful scientific fact, deliberately implied by some simple but pithy words:

-*Lo! Those who disbelieve Our revelations, We shall expose them to the fire. As often as their skins are consumed, We shall exchange them for fresh skins that they may taste the torment. Lo! Allah is ever Mighty, Wise*[1].

Chapter4: verse56

To provide a better understanding, we present a brief summary regarding the skin and its relationship to the nerves and the sense of touch. Then, we will analyze the corresponding Quranic viewpoint in this regard.

Scientific Viewpoint

What is Skin?

The skin is composed of several tissue layers. Located under the dermis is the layer of very loose connective tissue that joins the skin to the body. This layer, not technically part of the skin, houses many fat cells. The skin's inner layer or dermis consists of connective tissue surrounding several types of specialized structures, including hair follicles, sebaceous glands, sweat glands, capillary blood vessels, and nerves.

The skin's outer layer or epidermis, which is composed of stratified squamous cells, active cell division constantly occurs. The new cells push from deep within the layer toward the outside. The new cells continually replenish the epidermis's outermost layer, the stratum corneum, which consists of hardened, dead cells that are continually being sloughed off the body.

The skin's major tasks are to form a protective layer over the body to prevent injury and disease, keep moisture in the body (water retention), convert vitamin D to regulate body temperature and excrete wastes. The outer layer of skin is the epidermis. The inner layer is called the dermis. It contains hair follicles, nails, *nerves* (the body's sense of touch), sweat and oil glands and blood vessels.

Skin is the body's largest organ. An average square inch of skin contains 2,800 openings for sweat and oil glands, 72 feet of nerves, hundreds of pain and sense receptors and 15 feet of blood vessels. And skin is the body's most important barrier to infection. The new 3M(TM) Skin Health(SM) Program is designed for doctors and nurses to help their patients maintain healthy skin[2].

Quranic Viewpoint

Having considered the above scientific explanation, we can easily arrive at a conclusion that to experience pain (or any other sensations), it is necessary to have skin.

In a comparative analysis, we deduce many of the described functions of the skin from the mentioned corresponding verse in the Quran. The word "Leyazuqu" meaning "In order to make them taste" implicitly clarifies the necessity of existence of skin in order to feel the sensation. This is because after consumption (burning) of the unbelievers' skin in the fire, Allah replaces a new skin for them to taste (or to sense) the torment again and this happens repetitively to provide an eternal punishment for the unbelievers. Namely, since sensing the pain is at the skin, the doomed unbelievers will get fresh skin for their burnt out skin to taste it again.

Miracle in the Quran[3]

A scientist became Muslim after reading about these facts in the Holy Quran. This man declared the Islamic creed (*Shahaadah*) and became Muslim during the Eighth Saudi Medical Conference, which was convened in Riyadh. He is Professor Tejatat Tejasen, Chairman of the Department of Anatomy at Chiang Mai University in Thailand. He was previously the Dean of the Faculty of Medicine at the same university.

Sheik Abdul Majid Zendani, a Professor of Islamic Studies in King Abdulaziz University in Jeddah, Saudi Arabia says: "We presented to Professor Tejasen some Quranic verses and Prophetic Ahadeeth (narrations), which deal with his specialization in the field of anatomy.

Professor Tejasen, at first supposed that very accurate descriptions of embryonic developmental stages exist in their Buddhist books. We told him that we were very anxious and interested to see those descriptions and learn about these books. A year later, Professor Tejasen came to King Abdul Aziz University as an outside examiner. He was reminded of the statement that he made one year before. He apologized and said that he in fact had made that statement without researching the matter fully. Actually, when he checked the Buddhist resources he found that they contained nothing of relevance to the subject.

Upon this, he was presented a lecture written by Professor Keith Moore about the compatibility of modern embryology with what is contained in the Quran and the Sunnah. After reviewing it Professor Tejasen was asked if he knew of Professor Keith Moore's reputation and accomplishments. He knew Professor Moore as one of the most world-renowned scientists in the field of embryology.

Professor Tejasen's astonishment was observable after reviewing the article according to sources. He was presented with several questions in his own field of specialization, dermatology. One of the questions pertained to modern discoveries in dermatology about the sensory characteristics of the skin. Dr. Tejasen responded: Yes if the burn is deep.

Dr. Tejasen was told: "You will be interested to know that in this book, the Holy Book - the Quran was revealed 1400 years ago. This quote, which pertains to the moment of punishment of the unbelievers by the fire of Hell and it states that when their skin is destroyed, Allah makes another skin for them so that they perceive the punishment by a fire, indicating knowledge about the nerve endings in the skin, and the verse is as follows":

-Those who reject our signs, We shall soon cast into the fire. As often as their skins are roasted through, We shall change them for fresh skins, that they may taste the chastisement. Truly Allah is Exalted in Power, Wise.

Professor Tejasen was then asked: "Do you agree that this is a reference to the importance of the nerve endings in the skin in sensation, 1400 years ago?" Dr. Tejasen responded: "Yes I agree".

Further explanation of this expert from the Quran reveals that this knowledge about sensation was supported by modern evidence. It states in the Quran that if somebody does sins, then he will be punished by burning his skin and then Allah puts a new skin on him, covers him, to make him know that the taste is painful again. It means that they knew it many years ago that the receptor of pain sensation must be on the skin, so they put a new skin on.

The skin is the center of sensitivity to burns. Thus, if the skin is completely burnt by fire, it looses its sensitivity. It is for this reason that Allah will punish the unbelievers on the Day of Judgment by returning to them their skins time after time, as He, the Exalted and Glorified, said in the Quran:

-Those who reject our signs, We shall soon cast into the fire. As often as their skins are roasted through, We shall change them for fresh skins, that they may taste the chastisement. Truly Allah is Exalted in Power, Wise.

Professor Tejasen was presented with yet another question: 'Is it possible that these verses came to the Prophet Muhammed, (Peace Be Upon Him), from a human source?'

Professor Tejasen conceded that they could have never come from any human source 14 hundred years ago. He remained unable to identify the source of that knowledge and from where could Muhammed have possibly received it.

It was told to him that this is 'From Allah, the Most Glorified and Most Exalted.' Then he asked: 'But who is Allah?'

He was replied: "He is the Creator of all that is in existence.' If you find wisdom then it is because it comes only from the one Who is Most Wise. If you find knowledge in the making of this universe, it is because the universe is the creation of the One Who has all the knowledge. If you find perfection in the composition of these creations, then it is proof to you that it is the creation of the One Who Knows Best; and if you find mercy, then this bears witness to the fact that it is the creation of the One Who is Most Merciful. In the same way, if you perceive creation as belonging to one unified order and tied together firmly, then this is proof that it is the creation of the Only Creator, May He be Glorified and Exalted.

Professor Tejasen agreed with what we said to him. He returned to his country where he delivered several lectures about his new knowledge and discoveries. We were informed that five of his students converted to Islam as a result of these lectures. Then at the Eighth Saudi Medical Conference held in Riyadh, Professor Tejasen attended a series of lectures on Medical signs in the Quran and Sunnah.

Professor Tejasen spent four days with several scholars, Muslims and non-Muslims, talking about this phenomenon in the Quran and the Sunnah. At the end of those sessions Professor Tejasen stood up and said: "In the last three years I became interested in the Quran, which Shaykh Abdul-Majeed Az-Zindani gave me. Last year, I got Professor Keith Moore's latest script from the shaykh. He asked me to translate it into the Thai language and to give a few lectures to the Muslims in Thailand. I have fulfilled his request. You can see that in the videotape that I have given to the shaykh as a gift. From my studies and from what I have learned throughout this conference, I believe that everything that has been recorded in the Quran 1400 years ago must be the truth, which can be proven by scientific means.

He continued: "Since the Prophet Muhammed could neither read nor write, Muhammed must be a messenger who relayed this truth which was revealed to him as enlightenment by the One Who is an eligible Creator. This Creator must be Allah, or God. Therefore, I think this is the time to say 'Laa ilaaha illallah' (that there is no god to worship except Allah), 'Muhammed Rasool Allah' (Muhammed is the messenger of Allah...)

He continued on: "I have not only learned from the scientific knowledge in the conference, but also the great chance of meeting many new scientists and making many new friends among the participants. The most precious thing that I have gained by coming to this conference is 'La ilaaha illallah, Muhammed Rasool Allah', and to have become a Muslim."

The truth verily comes from Allah who said in the Quran:

-And those to whom knowledge has come see that the (revelation) sent down to thee from thy Lord - that is the truth, and that it guides to the path of the Exalted (in Might), worthy of all praise.
Chapter34: verse6

References
1. Holy Quran translated by Marmaduke Pickthall, George Allen and Unwin Ltd., London, Fifth Edition, 1969.
2. Caduceus MCAT Review, a product of Scientia Inc.
3. http://www.islim-guide.com/.

Cerebrum

كَلَّا لَئِن لَّمْ يَنتَهِ لَنَسْفَعًۢا بِٱلنَّاصِيَةِ ۝ نَاصِيَةٍ كَاذِبَةٍ خَاطِئَةٍ ۝

The Quran through this verse describes the Lobe (forehead) as being a sinful liar:

-Let him beware! If he desist not We will drag him by the forelock, a lying sinful forelock[1].

Chapter96: verse15 and 16

    How could the lobe be a liar? Why wasn't any other part of the body described as being a liar or mistaken? Is the lobe or forehead responsible for such behavior as lying or truthfulness; doing right or wrong?

    If we look at the anterior portion of the brain, we will find the prefrontal area of the cerebrum. What does physiology tell us about the function of this area? A book entitled, Essentials of Anatomy and Physiology, says about this area:

    "The motivation and the foresight to plan and initiate movements occur in the anterior portion of the frontal lobes, the prefrontal area. This is a region of the association cortex..." The book also says: "In relation to its involvement in motivation, the prefrontal area is also thought to be the functional center for aggression…"

Scientific Viewpoint

Cerebrum

The cerebrum, which develops from the front portion of the forebrain, is the largest part of the mature brain. It consists of two large masses, called "cerebral hemispheres", which are almost mirror images of each other. They are connected by a deep bridge of nerve fibers called the "corpus callosum" and are separated by a layer called the "falx cerebri". The surface of the cerebrum is marked by numerous ridges or "convolutions", called "gyri", which are also separated by grooves. A shallow groove is called a "sulcus", and a very deep one is a "fissure". A "longitudinal" fissure separates the right and left hemispheres of the cerebrum, and a "transverse" fissure separates the cerebrum from the cerebellum.

Various sulci divide each hemisphere into "lobes" (sometimes called "poles"). The lobes are named for the skull bones under which they rest and are: A) the frontal lobe, B) the parietal lobe, C) the temporal lobe, D) the occipital lobe, and E) the insula.

The cerebrum is concerned with higher brain functions, interpreting sensory impulses and initiating muscle movements. It stores information and uses it to process reasoning. It also functions in determining intelligence and personality.

Anatomy of the forehead

Warwick and Williams (1973) studied the anatomy of the forehead and found it is formed of one of the bones of the skull called the (Frontal Bone). This in turn protects one of the lobes of the brain called "the Frontal Lobe". Thus we can say that the frontal lobe of the brain is that organ hidden behind the forehead.

The human brain contains five nervous centers; one is in the Pre-Frontal Cortex, which represents the bigger of the "Frontal Lobe". Its function is related to the individual's personality formation. It is also considered as a high center of concentration, thinking and memory and has a role in controlling human feelings in addition to its effect on initiative behavior and judgment. This Pre-Frontal Cortex is laid directly behind the forehead, i.e. it disappears in the depth of the forehead and directs human behavior that reflects the personality such as lying, truthfulness, and doing wrong. It also urges the person to take the initiative whether to do right or wrong and so on.

Scientific Signs in human behavior with regard to the lobe

In addition to previously stated information about the anatomy of the frontal lobe of the brain and its function, research presents reports of some cases of patients who had a disturbance in that area (Pre-Frontal Cortex) such as cancer, accidents or internal hemorrhage that accumulates around the frontal lobe and causes pressure on this area. This would affect human behavior. Instead of the person being considerate towards the others around him, he becomes indifferent, and loses his sense of responsibility.

Of these cases mentioned, there was a patient who was hit in a car accident and had a fracture in the front part of the skull, and a hemorrhage in the front lobe. After saving his life by removing this internal hemorrhage from inside the skull and outside the brain, he still had some hemorrhaging inside the front lobe to the size of 4 mm. It was responsible for a complete change in the patient's behavior. After six months, a complete absorption of that hemorrhage occurred, his condition improved, and he went back to his job.

Another cited case was of a patient (a lady) who had a benign tumor in the frontal part of the skull, which was pressing on the Frontal Lobe for a long time causing a change in her behavior. Her relatives thought it was due to her age, but after a successful operation, she improved after less than one week became conscious and now behaves normally[2].

The Lobe and the higher mental functions

For several years scientists thought that the frontal parts of the brain called Lobes, were silent areas and that they had an insignificant role in controlling bodily functions. The reason for this belief rests on the fact that cutting the fibrous nerves of the frontal lobe, did not lead to any change in the animals activity.

In the past 50 years, the function of the frontal lobe was found to be related to higher mental functions in man and animals. EEG studies, and studies of electrostatic functions of the organs, showed that patients and animals which were subject to damage of the frontal lobe usually suffer from a decrease in mental ability and in humans they may suffer from a decrease in more standards. These lobes are connected with higher mental operations and affect the functions of other parts of the brain such as thinking, and feelings as well[3].

Quranic Viewpoint

So, this area of the cerebrum is responsible for planning, motivating, and initiating good and sinful behavior, and is responsible for telling lies and speaking the truth. Thus, it is proper to describe the front of the head as lying and sinful when someone lies or commits a sin, as the Holy Quran said: ...a lying, sinful "naasiyyah" (front of the head)! Scientists have only discovered these functions of the prefrontal area in the last sixty years, according to Professor Keith Moore.

Scientific signs in the Quran and Sunnah with regard to the Lobe

This verse was revealed when Abou-Jahl threatened the Prophet for praying at the Ka'ba. So, Allah threatened to take Abou Jahl by his lying forehead (lying in deeds and sayings), for he deserves severe chastisement.

Based upon the previous facts about the function of the Pre-Frontal Cortex, which is directly behind the forehead, we now understand the relevance to the Quranic expression, which described it as being a liar. This is because it is the center controlling the individual's behavior, as much as it expresses the different aspects of his personality whether truthful or not.

Scientific signs with regard to the Lobe

Allah the All Mighty says in the Quran "Nay! If he (Abu Jahl) ceases not, We will catch him by the forehead. A lying, sinful forehead: "96:15-16". Many interpreters of the Quran have indicated that this means the foreheads. The forehead actually refers to what is behind the bones of the forehead. All human behavior initiates in the frontal lobe. Lying, telling the truth or otherwise originate here and then the message passes to the speech organs or other related organs.
This is the pathway with other sins as well. They are planned in the frontal lobes before they are carried out by the eye, hand, sex organs etc. There is a Tradition of the Prophet (Peace Be Upon Him) that says: " Any one who is in distress or grief (can make this supplication): O' Allah I'm your slave, the son of your slave, my forehead is in Thy hand, your creed is executed, You are Just in your judgment. I ask Thee with every Attribute you have chosen for yourself or taught any of your creatures or mentioned in your book".

This Tradition shows that decree is in the hands of Allah, as much as the forehead is also in His hands. It signifies that the forehead has a very significant role in the direction of human behavior. This knowledge of the lobe, was far from imaginable at that era (14 centuries ago). Consequently those statements that were revealed to the Prophet in the Quran imply an awareness of the function of the frontal lobe which was not known at that time and was not even known except after deep and accurate studies of the physiology of organs (brain included). Isn't this further evidence that these revelations are the words of Allah? "Say:

-It (this Quran) has been sent down by Him (Allah) Who knows the secret of the heavens and the earth. Truly, He is Oft Forgiving, Most Merciful.

Chapter25: verse6

Scientists have only discovered these functions of the prefrontal area in the previous 60 years [4].

Could it be a coincidence that all this recently discovered scientific information mentioned in the Quran, was revealed 14 centuries ago? We may feel compelled to bear witness that Holy Quran is wholly the word of God, Almighty and an everlasting miracle indeed.

References
1. Holy Quran translated by Abdullah Yusuf Ali, DarAl Arabia Publication, Beirut, Lebanon, 1968.
2. By: Dr. Yehia Naser Khawagi, Dr. Khalaf El-Matari and Dr. Nizar El-Ameri.
3. By: Prof. Keith L. Moore and Sheikh Abdul-Majid Al Zindani.
4. By: Dr. Ahmad Kamal Mostafa, Dr. Mostafa An-Naggar and Dr. Faisal Zaher.

Further Sources
- Essentials of Anatomy and Physiology, Seeley and others, p. 211, also see The Human Nervous System, Noback and others, pp. 410-411.
- Al-E'jaz al-Elmy fee al-Naseyah (The Scientific Miracles in the Front of the Head), Moore and others, p. 41.
- http://www.islam-guide.com/.
- http://www.innerbody.com/text/nerv41.html.

Fingerprints

An amazing miracle is manifested in these verses:

-Does man think We shall not gather his bones?
-Yeah! We are able to make complete his very fingertips[1].
Chapter75, verses3 and 4

    Why has Allah chosen the human fingertips as an example among other various body limbs and organs? It is because human body's other organs like eyes, nose, ears, etc may possibly be similar among two people, but a human's fingertips are specifically individualized and fail to match anyone else' fingertips at all.

    Therefore, Allah presents His ability of making complete the human's fingertips in the doomsday as a sign for His omnipotence to resurrect the whole human body as exactly what he had been in his worldly life. This is in order to being accountable before Allah for what he did.

    Of course, this characteristic was not discovered until the 19th century, namely, twelve and a half centuries after the revelation of Holy Quran. In England, in 1884, fingerprints were formally used for ratification, since everyone's fingertips are made up of 3 kinds of lines:

1-Half circle lines

2-Parallel lines
3-Concentric circular lines

Also a 4th kind of lines is seen which consists of all shapes other than the mentioned types; and are called compound lines.

The noticeable point is that these individual shapes do not change throughout one's whole life; and the difference between the fingerprints among the people is very well distinguishable and unique to that person.

The History of Fingerprints

Pre-historic picture writing of a hand with ridge patterns have been discovered in Nova Scotia. In ancient Babylon, fingerprints were used on clay tablets for business transactions. In ancient China, thumbprints were found on clay seals. In 14th century Iran, various official government papers had fingerprints (impressions), and one government official, a doctor, observed that no two fingerprints were exactly alike.

In 1686, Marcello Malpighi, a professor of anatomy at the University of Bologna, noted in his treaties; ridges, spirals and loops in fingerprints. He made no mention of their value as a tool for individual identification. A layer of skin was named after him; "Malpighi" layer, which is approximately 1.8mm thick.

John Evangelist Purkinji - 1823 in 1923, John Evangelist Purkinji, a professor of anatomy at the University of Breslau, published his thesis discussing 9 fingerprint patterns, but he too made no mention of the value of fingerprints for personal identification.

The English first began using fingerprints in July of 1858, when Sir William Herschel, Chief Magistrate of the Hooghly district in Jungipoor, India, first used fingerprints on native contracts.

On a whim, and with no thought toward personal identification, Herschel had Rajyadhar Konai, a local businessman impress his hand print on the back of a contract. The idea was merely ". . . to frighten [him] out of all thought of repudiating his signature." The native was suitably impressed, and Herschel made a habit of requiring palm prints--and later, simply the prints of the right Index and Middle fingers--on every contract made with the locals. Personal contact with the document, they believed, made the contract more binding than if they simply signed it.

Thus, the first wide-scale, modern-day use of fingerprints was predicated, not upon scientific evidence, but upon superstitious beliefs. As his fingerprint collection grew, however, Herschel began to note that the inked impressions could, indeed, prove or disprove identity.

While his experience with fingerprinting was admittedly limited, Sir Herschel's private conviction that all fingerprints were unique to the individual, as well as permanent throughout that individual's life inspired him to expand their use.

Dr. Henry Faulds - 1880 during the 1870's, Dr. Henry Faulds, the British Surgeon-Superintendent of Tsukiji Hospital in Tokyo, Japan, took up the study of "skin-furrows" after noticing finger marks on specimens of "prehistoric" pottery.

A learned and industrious man, Dr. Faulds not only recognized the importance of fingerprints as a means of identification, but devised a method of classification as well. In 1880, Faulds forwarded an explanation of his classification system and a sample of the forms he had designed for recording inked impressions, to Sir Charles Darwin. Darwin, in advanced age and ill health, informed Dr. Faulds that he could be of no assistance to him, but promised to pass the materials on to his cousin, Francis Galton.

Also in 1880, Dr. Faulds published an article in the Scientific Journal, "Nautre" (nature). He discussed fingerprints as a means of personal identification, and the use of printers ink as a method for obtaining such fingerprints. He is also credited with the first fingerprint identification of a greasy fingerprint left on an alcohol bottle.

In 1882, Gilbert Thompson of the U.S. Geological Survey in New Mexico used his own fingerprints on a document to prevent forgery. This is the first known use of fingerprints in the United States.

Mark Twain (Samuel L. Clemens) in his book, "Life on the Mississippi"(1883), a murderer was identified by the use of fingerprint identification. In a later book, "Pudd'n Head Wilson", there was a dramatic court trial on fingerprint identification. A more recent movie was made from this book.

In 1888 Sir Francis Galton, a British anthropologist and a cousin of Charles Darwin, began his observations of fingerprints as a means of identification in the 1880's. In 1892, he published his book, "Fingerprints", establishing the individuality and permanence of fingerprints.

The book included the first classification system for fingerprints. Galton's primary interest in fingerprints was as an aid in determining heredity and racial background. While he soon discovered that fingerprints offered no firm clues to an individual's intelligence or genetic history, he was able to scientifically prove what Herschel and Faulds already suspected: that fingerprints do not change over the course of an individual's lifetime, and that no two fingerprints are exactly the same.

According to his calculations, the odds of two individual fingerprints being the same were 1 in 64 billion. Galton identified the characteristics by which fingerprints can be identified. These same characteristics (minutia) are basically still in use today, and are often referred to as Galton's Details.

In 1891, Juan Vucetich, an Argentine Police Official, began the first fingerprint files based on Galton pattern types. At first, Vucetich included the Bertillon system with the files. (See Bertillon below) In 1892, Juan Vucetich made the first criminal fingerprint identification. He was able to identify a woman by the name of Rojas, who had murdered her two sons, and cut her own throat in an attempt to place blame on another. Her bloody print was left on a door post, proving her identity as the murderer.

The year 1901 is of great importance to the history of fingerprinting, introduction of fingerprints for criminal identification in England and Wales. Using Galton's observations and revised by Sir Edward Richard Henry. Thus began the Henry Classification System, used even today in all English speaking countries.

In 1902 the first systematic use of fingerprints in the US by the New York Civil Service Commission for testing began. Dr. Henry P. De Forrest pioneers US fingerprinting for this purpose.

In the next year, 1903 The New York State Prison system began the first systematic use of fingerprints in US for criminals.

This followed in 1904, when the use of fingerprints began in Leavenworth State Penitentiary in Kansas, and the St. Louis Police Department. A Sergeant from Scotland Yard who had been on duty at the St. Louis Exposition guarding the British Display assisted them.

Then in 1905 history saw the use of fingerprints for the U.S. Army. Two years later the US Navy started, and joining the next year was the Marine Corp.

During the next 25 years more and more law enforcement agencies join in the use of fingerprints as a means of personal identification. Many of these agencies began sending copies of their fingerprint cards to the National Bureau of Criminal Identification, established by the International Association of Police Chiefs.

It was in 1918 when Edmond Locard wrote that if 12 points (Galton's Details) were the same between two fingerprints, it would suffice as a positive identification. This is where the often quoted (12 points) originated.

There is "NO" required number of points necessary for a positive identification. Some countries have set their own standards, which do include a minimum number of points, but not in the United States.

In 1924, an act of congress established the Identification Division of the FBI. The National Bureau and Leavenworth consolidated to form the nucleus of the FBI fingerprint files. By 1946, the FBI had processed 100 million fingerprint cards in manually maintained files; and by 1971, 200 million cards. With the introduction of AFIS technology, the files splitting into computerized criminal files and manually maintained civil files.

Many of the manual files were duplicates though, the records actually represented somewhere in the neighborhood of 25 to 30 million criminals, and an unknown number of individuals in the civil files. In 1999, the FBI plans to stop using paper fingerprint cards (at least for the newly arriving civil fingerprints) inside their new Integrated AFIS (IAFIS) site at Clarksburg, WV. IAFIS will initially have individual computerized fingerprint records for approximately 33 million criminals.

Old paper fingerprint cards for the civil files are still manually maintained in a warehouse facility (rented shopping center space) in Fairmont, WV. Since the Gulf War, most military fingerprint enlistment cards received have been filed only alphabetically by name... the FBI hopes to someday classify and file these cards so they can be of value for unknown casualty (or amnesiac) identification (when no passenger/victim list from a flight, etc., is known).

Why Fingerprint Identification?

Fingerprints offer an infallible means of personal identification. That is the essential explanation for their having supplanted other methods of establishing the identities of criminals reluctant to admit previous arrests. Other personal characteristics change - fingerprints do not.

In earlier civilizations, branding and even maiming were used to mark the criminal for what he was. The thief was deprived of the hand, which committed the thievery. The Romans employed the tattoo needle to identify and prevent desertion of mercenary soldiers. More recently, law enforcement officers with extraordinary visual memories, so-called "camera eyes," identified old offenders by sight. Photography lessened the burden on memory but was not the answer to the criminal identification problem. Personal appearances change.

Around 1870, a French anthropologist devised a system to measure and record the dimensions of certain bony parts of the body. These measurements reduced to a formula, which, theoretically, would apply only to one person and would not change during his/her adult life. This Bertillon system, named after its inventor, Alphonse Bertillon, was generally accepted for thirty years.

The Bertillon system never recovered from the events of 1903, when a man named Will West was sentenced to the U.S. Penitentiary at Leavenworth, Kansas. You see, there was already a prisoner at the penitentiary at the time, whose Bertillon measurements were nearly exact, and his name was William West. Upon an investigation, there were indeed two men. They looked exactly alike, but were allegedly not related. Their names were Will and William West respectively. Their Bertillon measurements were close enough to identify them as the same person. However, a fingerprint comparison quickly and correctly identified them as two different people. The West men were apparently identical twin brothers per indications in later discovered prison records citing correspondence from the same immediate family relatives.

Scientific Viewpoint

Every living organism consists of tiny compartments called cells. These cells are basic structural and functional units of living organisms. Externally these cells are bound by a cell membrane. Inside the membrane, cell organelles are present each disposing its specific function.

The center of the cell is occupied by the nucleus. Nucleus is master organelle of the cell as it guides the cellular function. Inside the nucleus are packed thread-like structures called chromosomes. These chromosomes play a significant role in the transmission of characters from parents to the offspring.

George Mendle, an Austrian biologist (1822-1884) was the first to reveal that chromosomes contained certain mysterious elements responsible for the transmission of characters. Mendle named these elements as factors. Modern research in genetics has enabled us to know more about the actual mechanism behind the transmission of character from parents to the offspring.

Nowadays it is well established that chromosomes are made of certain chemical substances, which include:

1. Two types of nucleic acids-DNA and RNA
2. Protein.

These three substances, DNA, RNA, and proteins are thus responsible for storing and utilizing the vast amount of genetic information and for transmitting the same from one generation to other.

DNA is the hereditary material exercising the main genetic control. RNA plays a complementary role in the process leading to protein synthesis and Proteins act as organic catalysts (enzymes) to bring about the expression of specific traits. DNA is the primary genetic material. It is confirmed by the following facts, all of which do not hold for the other chromosome components (RNA and proteins).

The functional unit segment of DNA consisting of several sub units (nucleotide pairs) is commonly referred to as gene. Most significantly, no two individuals have the same genetic make up in this universe (exception may be found in monozygotic twins).

It should be clear that every individual has his own genetic make up which governs its phenotypic characteristics i.e. external make up and further it is also clear that no two individuals have similar genome, so individuals differ both genetically and phenotypically. And it also should be pointed out that when a human body dies, it may undergo disintegration under chemical and microbial action within the soil.

Quranic Viewpoint

Elements, which make up the body, are not destroyed and under the creative powers of Allah the elements assemble again as testified by the Quran:

-We already know how much of them the earth takes away; with us is the record guarding.

Chapter50: verse4

And this reconstruction will follow entirely the structure and contents of the body as it previously existed and would surely be based entirely upon its DNA content. The DNA content would express itself with such a precision and accuracy that it would even determine the structure of very tips of fingers as described by the Quran quoted earlier.

This is the point, which is referred to in Chapter75, verses3 and 4 really a scientific miracle of the Quran described some fourteen hundred years back:

-Does man think We shall not gather his bones?
-Yeah! We are able to make complete his very fingertips.

All this defies the human intelligence. One can say nothing except that the source of Holy Quran is Allah and is really a guiding book for everyone.

Holy Quran is not a book of some prosaic words. Every speech of it has its own purport and each verse of it represents a fact.

References
1. Holy Quran translated by M.H. Shakir, Ansarian Publication, Qum, the Islamic Republic of Iran, 1993.

Scientific Sources
- Wonders of Quran, By Nasir Hussein Peerzadah, http://www.islamicvoice.com/.
- http://www.onin.com/fp/fphistory.html/.

Therapeutic Properties of Honey

وَأَوْحَىٰ رَبُّكَ إِلَى ٱلنَّحْلِ أَنِ ٱتَّخِذِي مِنَ ٱلْجِبَالِ بُيُوتًا وَمِنَ ٱلشَّجَرِ وَمِمَّا يَعْرِشُونَ ﴿٦٨﴾

ثُمَّ كُلِي مِن كُلِّ ٱلثَّمَرَٰتِ فَٱسْلُكِي سُبُلَ رَبِّكِ ذُلُلًا يَخْرُجُ مِنۢ بُطُونِهَا شَرَابٌ مُّخْتَلِفٌ أَلْوَٰنُهُۥ فِيهِ شِفَآءٌ لِّلنَّاسِ إِنَّ فِى ذَٰلِكَ لَءَايَةً لِّقَوْمٍ يَتَفَكَّرُونَ ﴿٦٩﴾

At the present time, the medical scientists have discovered that honey is composed of 25-40% glucose and fructose, which have numerous therapeutic effects in healing many diseases. Fourteen hundred years ago, Holy Quran vividly announced that honey is "healing for mankind":

-And thy Lord inspired the bee saying: Choose thou habitations in the hills and in the trees and in that which they thatch.
-Then eat of all fruits and follow the ways of thy Lord, made smooth (for thee), there cometh forth from their bellies a drink diverse of hues, wherein is healing for mankind. Lo! Herein is indeed a portent for people who reflect[1].

Chapter16: verse68 and 69

Let us see what science has to say about the healing property of honey in addition to being a sweet drink for mankind.

Scientific Viewpoint

Honey, sweet, thick, supersaturated sugar solution manufactured by bees to feed their larvae and provide subsistence for their hive in winter. It is produced when the nectar of flowers is ingested by worker bees and converted to honey in special sacs in their esophagi. It is stored and aged in combs in their hives.

Honey is an important constituent of the diet of many animals, such as bears and badgers, and is put to many uses by humans as well. Other insects, such as the honey-ant and various aphids, manufacture a honey-like substance from flowers, from the honeydew of plants, or from the sweet secretions elaborated by other insects.

Honey is composed of fructose, glucose, and water, in varying proportions; it also contains several enzymes and oils. The color and flavor depend on the age of the honey and on the source of the nectar. Light-colored honeys are usually of higher quality than darker honeys; white honey derives from the Californian white sage, Salvia apiana. Bees make other high-grade honeys from orange blossoms, clover, and alfalfa; a well-known, poorer-grade honey drives from buckwheat.

Honey has a fuel value of about 3307 cal/kg (about 1520 cal/lb). It readily picks up moisture from the air and is consequently used as a moistening agent for tobacco and in baking. Glucose crystallizes out of honey on standing at room temperature, leaving an uncrystallized layer of dissolved fructose. Honey to be marketed is usually heated by special processes to about 66° C (about 150° F) to dissolve the crystals and is poured into containers that are then sealed to prevent crystallization. The fructose in crystallized honey ferments readily at about 16° C (about 60° F). Fermented honey is used to make honey wine or mead[2].

According to the scientific investigations, honey has therapeutic effects in curing the following diseases:

Diabetes
Digestive Diseases
- Flatulence
-The Peptic and Intestine Ulcer and Inflammation
-Constipation
Respiratory Diseases
-Tuberculosis
-Lung Inflammation
Gynecopathy

-Vomits During Pregnancy
- Inflammation of Vagina
Ear, Pharynx and Nose Diseases
-Flu
-The Hidden Inflammation of Respiratory Channels: Pharyngitis, Rhinitis.
The Skin Diseases
-Skin Ulcers
-Pustules [3]

Quranic Viewpoint

The Quran through these verses explains the process of honey production by the bees in the nature and its healing properties:

-And thy Lord inspired the bee saying: Choose thou habitations in the hills and in the trees and in that which they thatch.
-Then eat of all fruits and follow the ways of thy Lord, made smooth (for thee), there cometh forth from their bellies a drink diverse of hues, wherein is healing for mankind. Lo! Herein is indeed a portent for people who reflect.

Chapter16: verse68 and 69

We present you Dr. Gary Miller's point of view regarding the therapeutic effects of honey, which is emphatically mentioned in the Quran:
"Various historical sources state that the Prophet gave some advice about health and hygiene. ...The Quran is a divine revelation, and as such, all information in it, is of divine origin. Allah revealed the Quran from Himself. It is the words of Allah, which existed before creation, and thus nothing can be added, subtracted or altered. In essence, the Quran existed and was complete before the creation of Prophet Muhammed (Peace Be Upon Him), so it could not possibly contain any of the Prophet's own words or advice. In fact, the Quran only mentions one item dealing with medical treatment, and it is not in dispute by anyone. It states that in honey there is healing. And certainly, I do not think that there is anyone who will argue with that" [4].

How could this wonderful information about the origin and the process of producing honey have existed in such a precise way in a book of 1400 years ago? Did the prophet of Islam, Muhammed major in medical sciences to know about the healing properties of honey? He was an unlettered man who lived in the Arabian Peninsula, with the barren deserts covering it and where there are not many plants, flowers and a convenient nature for the bees to live in and produce honey! True, as the verse says, *"Herein is indeed a portent for people who reflect"*. For those who have knowledge, Holy Quran is no less than a revelation.

References
1. Holy Quran translated by Marmaduke Pickthall, George Allen and Unwin Ltd., London, Fifth Edition 1969.
2. Honey, Encarta Encyclopedia 2001 http://encarta.msn.com/, 2001 Microsoft Corporation.
3. Quran and the New Medicine (in Persian language), By Doctor M. Noori Shadkaam, Dar Al-kotob Al-Islamiah, Teheran, 1999.
4. The Amazing Quran, by Dr. Gary Miller, http//:www.users.erols.com.

Menstruation

وَيَسْـَٔلُونَكَ عَنِ ٱلْمَحِيضِ قُلْ هُوَ أَذًى فَٱعْتَزِلُوا۟ ٱلنِّسَآءَ فِى ٱلْمَحِيضِ وَلَا تَقْرَبُوهُنَّ حَتَّىٰ يَطْهُرْنَ فَإِذَا تَطَهَّرْنَ فَأْتُوهُنَّ مِنْ حَيْثُ أَمَرَكُمُ ٱللَّهُ إِنَّ ٱللَّهَ يُحِبُّ ٱلتَّوَّٰبِينَ وَيُحِبُّ ٱلْمُتَطَهِّرِينَ ۝

Islam as the perfect religion, which gives prominence to all human aspects, offers the most complete guidelines to humanity leading him to the best and most upright lifestyle.

One of these human aspects is the personal hygiene, which is highly advised in Quran and Hadeeth (Islamic Narration). The following verse is an evident example on this issue. Allah, the Almighty says: (He) Loveth those who have a care for cleanness:

-*They question thee (O Muhammed) concerning menstruation. Say: "It is an illness, so let women alone at such time and go not in unto them till they are cleansed. And when they purified themselves then go in unto them as Allah hath enjoined upon you. Truly, Allah loveth those who turn unto Him and loveth those who have a care for cleanness"*[1].

Chapter2: verse222

This verse strictly orders men to abstain from sexual intercourse with women at the time of their menstruation. We are going to offer you some newly discovered reasons for this Quranic order from the standpoint of science.

Scientific Viewpoint

Menstruation

Menstruation is periodic vaginal discharge in humans and other mammals, consisting of blood and cells shed from the endometrium, or lining of the uterus. Menstruation accompanies a woman's childbearing years, usually beginning between the ages of 10 and 16, at puberty, and most often ceasing between the ages of 45 and 50, at menopause.

Menstruation is part of the process that prepares a woman for pregnancy. Each month the lining of the uterus thickens. If pregnancy does not occur, this lining breaks down and discharges through the vagina. The three to seven days that menstruation lasts is the menstrual period.

In most women the menstrual cycle is about 28 days, but it can vary considerably even from one month to another. Hormones in the blood stimulate the ovaries and initiate the cycle. (The ovaries are two female organs that produce ova, or eggs). Each month, hormones cause an egg in one of the two ovaries to mature (to become capable of being fertilized and develop into a fetus).

The ovaries also produce hormones of their own, primarily estrogen, which cause the endometrium to thicken. About midway through the menstrual cycle, 14 to 15 days before the next period, the ovary releases the mature egg in a process called ovulation. The egg passes through the Fallopian tube to the uterus. If the egg unites with a sperm on its way to the uterus, fertilization occurs and pregnancy ensues.

The three to five days the egg takes to reach, the uterus after release by the ovary is the woman's fertile period. If fertilization does occur, the fertilized egg attaches itself to the enriched uterine lining and pregnancy continues. Menstruation does not occur during pregnancy and a missed period is often the first indication of pregnancy a woman notices (see Pregnancy and Childbirth). If fertilization does not occur, the lining of the uterus does not receive the hormones it needs to continue the thickening process. Thus, the uterine lining breaks down and discharges from the body during menstruation.

Many women experience premenstrual discomfort. Tenderness of the breasts and a tendency to retain fluid (bloat) are common one to seven days before each period.

Some women also experience a condition called premenstrual syndrome (PMS), characterized by headaches, irritability, nervousness, fatigue, crying spells, and depression with no apparent cause. A few women also experience menstrual cramps (dysmenorrhea) during the first day or two of the period. Although premenstrual symptoms and discomfort during menstruation were once thought to be of psychological origin, research now indicates that hormonal and chemical changes are responsible. New medications are effective in treating these problems [2].

Harmful Effects of Coitus on Women at the Time of Menstruation

The spiral arteries, which bleed the womb's internal tissue (endometrium) constrict in the time of women's monthly menstrual period. Consequently, the bleeding to endometrium is disordered and causes the lining of the uterus to collapse due to the decrease of female hormones (estrogen and progesterone). This collapse of endometrium is accompanied by discharging an amount of blood, which is called the menstruation.

The collapse of this tissue prepares a ready environment for putrefaction inside the womb and ultimately, inflammation in the abdomen if being contaminated by pathogenic bacterium.

One of the factors, which accelerates and ease transferring the pathogenic bacterium to the collapsed tissue, is coitus in the time of menstruation. Therefore, doctors strictly recommend no sexual intercourse with women in this circumstance.

The female hormones affect the vagina's muscles and causes them become relaxed and limp. This relaxation of muscles causes the easy and painless coitus. As we stated above, the female hormones decrease in the time of monthly menstruation. Consequently, the muscles are not relaxed. This is one of the factors, which brings about painful coitus in this situation.

In the time of monthly menstrual period, the vagina's secretion diminishes because of the decrease of female sex hormones. These secretions perform some functions:

A) To ease the coitus
B) To preserve the sperm and leading it to reach the womb.
C) Since these secretions contain the acidic pH, so it prevents the growth and reproduction of microbes inside the womb.

Diminishing of these secretions and the excretion of blood from the womb changes pH in vagina and womb, and provides a ready place for pathogenic bacterium to grow and consequently, the putrefaction in womb and vagina.

One of the effects of estrogen female hormone is the increase of sexual desire in women. For this reason, the women's sensuality increases in the middle of menstrual period (12th_ 16th days). The diminishing of estrogen sex hormone in the menstrual period causes the hatred of coitus and the decrease of women's sexual desire. Perhaps, it is for this reason that Allah addresses men in the verse and forbids them from sexual intercourse in this situation, since men mostly impose their desires upon women in this regard. It may be that the imposing a desire upon woman, which she dislikes, causes the decrease of affection between the two mates.

Also it is known that the coitus brings about the blood congestion in woman's genital organs, it increases the possibility of bleeding in menstrual period.

Lastly, because of menstrual bleeding, sex is unattractive and repulsive. Therefore, the sexual intercourse is not beneficial for the couple from the psychosexual viewpoint[3].

Quranic Viewpoint

However, the Quran prevents men from the coitus at the time of menstruation:

-They question thee (O Muhammed) concerning menstruation. Say: "It is an illness, so let women alone at such time and go not in unto them till they are cleansed. And when they purified themselves then go in unto them as Allah hath enjoined upon you. Truly, Allah loveth those who turn unto Him and loveth those who have a care for cleanness".

Chapter2: verse222

Fourteen centuries after the appearance of Quran in the peninsula of Arabia, the contemporary medical scientists have discovered that the coitus in this situation is seriously harmful for both men and women for the medical reasons mentioned above.

We never claim that the philosophy of what Allah has ordained in this regard is exactly what we stated above. However, we can explicitly claim that the medical discoveries are in accordance with this miraculous Quranic order.

References
1. Holy Quran translated by Marmaduke Pickthall, George Allen and Unwin Ltd., London, Fifth Edition 1969.
2. Contributed By: Michaela P. Richardson, BS. Chief, Office of Research Reporting, National Institute of Child Health and Human Development, National Institutes of Health, USA.
3. Quran and the New Medicine (in Persian language), By Doctor M. Noori Shadkaam, Dar Al-kotob Al-Islamiah, Teheran, 1999.

Geology in Holy Quran

- Geological Description of Mountains
- Lowest Land of Earth

Geological Description of Mountains

خَلَقَ ٱلسَّمَٰوَٰتِ بِغَيْرِ عَمَدٍ تَرَوْنَهَا وَأَلْقَىٰ فِى ٱلْأَرْضِ رَوَٰسِىَ أَن تَمِيدَ بِكُمْ وَبَثَّ فِيهَا مِن كُلِّ دَآبَّةٍ وَأَنزَلْنَا مِنَ ٱلسَّمَآءِ مَآءً فَأَنۢبَتْنَا فِيهَا مِن كُلِّ زَوْجٍ كَرِيمٍ ۝

أَلَمْ نَجْعَلِ ٱلْأَرْضَ مِهَٰدًا ۝

وَٱلْجِبَالَ أَوْتَادًا ۝

وَجَعَلْنَا فِيهَا رَوَٰسِىَ شَٰمِخَٰتٍ وَأَسْقَيْنَٰكُم مَّآءً فُرَاتًا ۝

وَٱلْجِبَالَ أَرْسَىٰهَا ۝

وَأَلْقَىٰ فِى ٱلْأَرْضِ رَوَٰسِىَ أَن تَمِيدَ بِكُمْ وَأَنْهَٰرًا وَسُبُلًا لَّعَلَّكُمْ تَهْتَدُونَ ۝

Whatever Quran describes about mountains may seem simple at the first glance, but the matter of fact is that each one indicates a wonderful geological fact.

The earth speeds through space at 29.8 km per second (18.5 miles per second) _75 times faster than Concorde. None of this movement makes us feel dizzy. We seem to stay still while everything else seems to move around us.

What is really the reason of this phenomenon? Why do we not feel dizzy in this moving cradle? We can find the answer in the first verse:

-...and (Allah) hath cast into the earth firm hills, so that it quake not with you[1].

Chapter31: verse10

True! The firm mountains prevent the earth from being shaken, so that we do not feel dizzy, although the earth is moving through space at such an incredible speed.

In this verse, the word "Rawaasia" is the plural of the noun "Rasiah" which literally means " Firm", "Stable" and "The Anchor"[2]. Needless to say, the function of anchor is "to fix" or "to prevent the boat from moving away"[3]. Interestingly enough, according to the scientists, the mountains play an important role in stabilizing the crust of the earth[4].

This characteristic of earth is also a wonderful similitude to a cradle in the other verse. A cradle is continuously shaken by the kind-hearted mother but the baby inside, not only does not feel dizzy but also this gentle motion dedicates the child a tranquility by which he or she goes to a nice sleep:

-Have We not made the earth as a cradle[5]?

Chapter78: verse6

This tranquility, which we feel in the cradle of the earth, is a direct effect of the mountains. Some other points in the verses above should also attract our full attention:

a) The mountains resemble the pegs, which hold the land (Because the pegs are utilized to prevent things from being scattered).

b) There is a relationship between the mountains and water supplies, since the subject of drinkable water is mentioned immediately after the word "mountains" in the other verse, presupposes a close relation between these two subjects.

Each of these points are all assumed to be miraculous and scientific aspect of Holy Quran, leads us to the majesty of this divine book and makes us acquainted with its deep contexts.

To achieve certainty regarding the authenticity of the miraculous aspects of the preceded verses, we have provided you the following scientific passage.

Scientific Viewpoint

Mountains, the Skeleton of Earth

Mountains act actually like the steel armor, which have surrounded the earth. Since they connect to each other in beneath, they form a strong network.

If the crust of the earth were formed of only a soft and fine layer like soil and sand, would be indeed affected by the moon's gravity and would cause the tides- like what happens in the seas- so, it would always be in disturbance, worse than the earthquakes. In that case, we would not feel this tranquility that we have now in our lives.

Further more, this firm crust and steel armor can control the inner pressurized molten materials, which want to rip up the earth's crust and get out of it. If there were not the mountains, the earth face would definitely seem differently.

According to these facts about the mountains, do they not actually act like the earth's pegs; and prevent it from being disturbed?

Mountains, the mighty claws

The atmosphere's accumulation exerts a pressure of 1 kg per centimeter. Its total weight comes to 6 million billion tons.

Now guess what would happen, if the earth spun at a speed of 30 kg/min, while the air around itself was stagnant. A great deal of heat energy would be generated by the collision of the air molecules with the earth's surface potentially burning anything on it. Fortunately, the earth's low lands and elevations, especially the mountains have solved this great problem. They claw at the air and move it gradually within themselves, like the clock's gears. Consequently, this thick layer of air circulates along with earth, especially adjacent to the surface. This is also a kind of tranquility, dedicated to mankind by the mountains.

Mountains, the regulators of the winds

As we know, the winds blow permanently, from the poles to the equator and from equator to the poles, because of the difference of temperatures between them.

If these giant barriers of mountains did not control these movements of the winds, the whole surface of earth, like desserts, would permanently be exposed to the annoying storms and killing tempests, so there would not be quietude on earth for human.

Mountains, the Natural Refrigerators

If there were not mountains, the flowing rivers would not exist on earth. As we know, in high altitudes, the air is much colder and the heat radiation is less than what is on earth. Therefore, a huge amount of water is stored up in the form of snow and natural refrigerators over the mountains and constitutes the permanent and reliable sources of water for the rivers, while melting gradually in the warm seasons.

If there were not mountains, the rains would wash out the earth's surface in the form of floodwater, head up for the seas immediately, and the dry lands burn, in the lack of water reservation [6].

Quranic Viewpoint

As we mentioned above, according to the geologists, the mountains function as the natural refrigerators in which a huge amount of water is stored up in the form of snow to supply water in the warm seasons.
Wonderfully, this is what we perceive in the Quran, considering a relation between mountains and providing the drinkable water and river streams:

-And placed therein high mountains and given you to drink sweet water therein[7].

Chapter77: verse27

Being cognizant that Muhammed (peace be upon him) had never seen a high mountain especially with snow in his whole life in the desert of Arabia, one might feel compelled to admit that these realities are the certain evident for the scientific miracles of the Quran.

Sheik Abdul Majid Zendani, a Professor of Islamic Studies in King Abdulaziz University in Jeddah, Saudi Arabia relates from his meeting with the professor Siaveda, one of the best-known marine geologists from Japan and also one of the most famous scientists in the world:

"...He differs from other scholars but at the same time, he is a representative of a group of scholars. Professor Siaveda's mind was filled with many distortions and suspicions about all religions. He is right about what he says with regard to all religions, except Islam, because Islam differs from all other religions that he spoke about.

When we met him, he said to us: "You scholars of religion all over the world should keep your mouths shut forever". The doctor most assuredly was a great skeptic of religion.

We replied: 'But why Professor? Why?' He said: Because when you speak you cause wars to break out between the people throughout the world. We asked him: 'Why NATO alliance and WARSAW Pact are accumulating such massive nuclear arsenals and nuclear weapons in space, sea, overland, and underground. Why this? Is it for religious reasons?'

He went silent then we said to him: 'At any rate we are aware of your attitude towards all religions, but since you do not know much about Islam, then you might as well listen to what we have to say.' So we asked him a number of questions in his area of specialization, and we also informed him of the Quranic verses and ahadeeth (Narration), which mention the phenomena he spoke of. One of these questions was about mountains and whether they were firmly rooted in the earth.

He replied and said: "The fundamental difference between continental mountains and the oceanic mountains lies in its material. Continental mountains are made essentially by sediments, whereas the oceanic mountains are made of volcanic rocks. Continental mountains were formed by compressional forces, whereas the oceanic mountains were formed by extensional forces. But the common denominator on both mountains is that they have roots to support the mountains."

He continued: "In the case of continental mountains, light-low density material from the mountain is extended down into the earth as a root. In the case of oceanic mountains, there is also light material supporting the mountain as a root, but in the case of oceanic mountain this material is not light because the composition is light, but it is hot, therefore expanded somewhat. But from the viewpoint of densities, they are doing the same job of supporting the mountains. Therefore, the function of the roots is to support the mountains according to the law of Archimedes".

Professor Siaveda described the shape of all mountains whether they are on land or in the sea as being in the shape of a wedge. Could anyone during the time of Prophet Muhammed (Peace Be upon Him) have known of the shape of these mountains? Could anyone imagine that the solid massive mountain that he sees before him actually extends deep into the earth and has a root as scientists assure? A large number of geography books when discussing mountains, only describe that part which is on the surface of the earth. This is because specialists in geology do not write them, but modern science informs us about it and Allah says in the Quran:

-and the mountains as pegs[8]?

Chapter78: verse7

We asked Professor Sievada whether the mountains have a function in establishing the crust of the earth. He said that this has not yet been discovered and established by scientists. In view of the reply we researched and inquired about this and we found out that many geologists gave the same reply, except for a few.

Among those few are the authors of this book, entitled "The Earth". This book is considered as a basic reference text in many universities throughout the world. One of the authors of this book is Frank Press. He is currently the President of the Academy of Sciences in USA. Previously, he was science advisor to former U.S. President Jimmy Carter. What does he say in his book? He illustrates the mountains in a wedge-like shape where the mountain itself is but a small part of the whole whose root is deeply entrenched in the ground. Dr. Press writes on the functions of the mountains and states that they play an important role in stabilizing the crust of the earth.

This is exactly how the Quran described mountains 14 centuries ago. Allah said:

-And the mountains Has He firmly fixed.

Chapter79: verse32

And He said:

-And the mountains as pegs.

Chapter78: verse7

And He also said:

-And he has set up on the earth mountains standing firm, lest it should shake with you.

Chapter16: verse15

Who could have informed Prophet Muhammed (Peace Be upon Him) about this? We asked Professor Siaveda this question: 'what is your opinion on what you have seen in the Quran and the Sunnah with regard to the secrets of the Universe, which scientists only discovered now?' His answer was: "I think it seems to me very, very mysterious, almost unbelievable. I really think if what you have said is true; the book is really a very remarkable book..."

Yes, what can scientists say no matter how skeptical? They cannot attribute the knowledge revealed to Prophet Muhammed (Peace Be upon Him) by Allah and contained in the Quran to human beings or to any scientific authority in our times, because all scientists were unaware of all these secrets. Moreover, not all humanity could have any explanation but to attribute that knowledge to some extra-terrestrial force. Yes, it is a revelation from Allah, which he sent to his servant, the unlettered Prophet Muhammed, whom Allah has made an everlasting sign escorting humanity until the last hour[9].

References

1. Holy Quran translated by Marmaduke Pickthall, George Allen and Unwin Ltd., London, Fifth Edition 1969.
2. Al-Mawrid, A Modern English-Arabic Dictionary, by Munir Ba'albaki, Dar El-Ilm Lil-Malayen, Beirut, 1986.
3. Cambridge International Dictionary of English, Cambridge University Press, 1995.
4. Earth, Professor Emeritus Frank Press and Siever, p. 435. Also see Earth Science, Tarbuck and Lutgens.
5. Holy Quran translated by Arthur J. Arberry, Ansarian Publishers, Qum, the Islamic Republic of Iran, 1999.
6. The Quran and the Last Prophet (in Persian language), by ayatollah Makarem Shirazi, Dar AL-Kotob Al-Islamiah, Qum, the Islamic Republic of Iran, 1996.
7. Holy Quran translated by Arthur J. Arberry, Ansarian Publishers, Qum, the Islamic Republic of Iran, 1999.
8. Holy Quran translated by Marmaduke Pickthall, George Allen and Unwin Ltd., London, Fifth Edition 1969.
9. Mountains in Holy Quran, http://www.islam-guide.com/.

Lowest Land of Earth

غُلِبَتِ ٱلرُّومُ ۝ فِىٓ أَدْنَى ٱلْأَرْضِ وَهُم مِّنۢ بَعْدِ غَلَبِهِمْ سَيَغْلِبُونَ ۝

Geology has recently revealed a great deal of information to us about the world we live in. We now know that the Mount Everest is the highest peak and the Dead Sea is the lowest point of land on earth, which is located between Jordan and Palestine close to Jerusalem. Fourteen hundred years ago, the Quran in chapter *'Romans'*, emphatically implied that this point is the lowest land of earth:

-*The Romans have been vanquished,*
-*In the lowest part of the land; and after their vanquishing they shall be the victors,*
-*In a few years...*

Chapter30: verse2, 3 and 4

The story of these verses refers to the battle, which occurred in the seventh century between the empire of Persia (Iran) and the empire of Rome in a place near Jerusalem, previously referred to as the lowest point. The verse says: ***"The Romans have been vanquished in the lowest part of the land".*** In other words, it implies that the Romans have been defeated (by Persians) in a place, which is the lowest land of earth. It also correctly prophesied the Romans' dominance over the Persians, which actually occurred in a few years later.

To achieve a better idea about this Quranic viewpoint, we have to go over the respective geological and historical standpoints in this regard. Their conformity with the pertinent Quranic statement would certainly impel us to believe in the authenticity of Holy Quran as a miraculous revelation from God, Almighty.

In 614 AD, five years after Bi'thaat (appointment of Muhammed as the prophet), a battle occurred between Iran (Persia) and Rome (Byzantine) in a place near Jerusalem. King Khosro Parviz, the emperor of Iran invaded Jerusalem and committed genocide and took the cross ascribed to Jesus Christ with himself.

Parviz continuously dominated the Romans until he conquered the Asia Minor (now Turkey), which was the greatest part of the empire of Byzantine in that time. In 617 AD, he advanced on the city of Constantine. Heraclius tried to flee. He dispatched the government properties to Africa but they were trapped by the Persian troops in Roman Sea and were called as the "unlooked-for and easy come treasure". The great ones of Byzantine held Heraclius, not letting him flee and made him fight with the Persians. Eventually, in 622 AD, he vanquished the troops of Parviz in the port of Isus, which was located in the south of Asia Minor and this was the beginning of Romans' victory over the Persians...[1]

Scientific Viewpoint

Dead Sea, the lowest point on earth

The Dead Sea, 417 m below sea level, is the lowest point on the surface of the earth. It is at the terminus of the Jordan River, has no outlet and is very rich in minerals. As the name suggests, the sea is devoid of life due to an extremely high content of salts and minerals.

It is these natural elements, which give the waters their curative powers, recognized since the days of Herod the Great, 2,000 years ago. They also provide the raw materials for the renowned Jordanian Dead Sea Bath Salts and cosmetic products, which are marketed worldwide.

Fifty miles long and 10 miles wide, it is fed mainly by the River Jordan to the north. The intense heat causes the water to evaporate, leaving high concentrations of salts and minerals.

Surrounded by arid hills, as devoid of life as the sea itself, it glistens under a burning sun with barely a ripple disturbing its surface. The rocks that meet its lapping edges become covered with a snow-like thick gleaming deposit of white salt. This extremely high concentration of salt is what gives its waters the renowned therapeutic qualities and its buoyancy. Because the salt content is eight times that of most world's oceans, you can float in the Dead Sea without even trying. Swimming in it is a truly unique experience not to be missed [2&3].

Quranic Viewpoint

The Quran geologically speaks of the lowest part of the earth, where a battle took place between the Persians and the Romans. Allah, may He be Exalted and Glorified, said in the Quran:

-The Romans have been defeated, in the lowest part of the land, but after defeat they will soon be victorious.

Chapter30: verse2, 3 and 4

The term "adna" in Arabic means both lowest and nearer [4]. The commentators of the Quran were of the opinion that "adnal-ardh" meant the nearest land to the Arabian Peninsula. However, the first meaning is also there; that is, the lowest land. In this way, the Glorious Quran gives several meanings in one word.

Sheik Abdul Majid Zendani, a Professor of Islamic Studies in King Abdulaziz University in Jeddah, Saudi Arabia says about his meeting with the professor Palmer:

"When we investigated the lowest part of the earth, we found that it was exactly the same spot that witnessed the battle in which the Romans were defeated. When we informed Professor Palmer about this, he contested that there were many other areas, which are lower than the one referred to in the Quranic verse. He gave examples and names of other areas in Europe and in the United States. We assured him that our information was verified and correct. He had with him a topographical globe that showed elevations and depressions. He said that it would be easy with that globe to ascertain, which was the lowest spot on earth. He turned the globe with his hands and focused his sign on the area near Jerusalem. To his astonishment, there was a small arrow sticking out towards that area with words: 'the lowest part on the face of the earth.'"

The Sheik continues in his description of the precipitating events: "Professor Palmer was quick to concede that our information was correct. He proceeded to speak, as you now see him with the globe, saying that this was actually the lowest part of the earth."

"Professor Palmer: It took place in the area of the Dead Sea which is up here and interestingly enough the labeling on the globe says "the world's lowest point". So it certainly is supported by the interpretation of that critical word."

Sheik Abdul Majid Zindani relates: "Professor Palmer was even more astonished when he found that the Quran talks about the past and describes how creation first began; how the earth and heavens were created; how the water gushed forth from the depth of the earth; how the mountains were anchored on land; how vegetation first began; how is earth today, describing the mountains, describes its phenomena, describes the changes on the surface of the earth as witnessed in the Arabian Peninsula. It even describes the future of the land of Arabs and the future of the whole earth. At this, Professor Palmer acknowledged that the Quran is such a wondrous Book, which describes the past, the present, and the future."

Like many other scientists, Professor Palmer was hesitant at first. After reflection he was quick at forthcoming with his opinions. In Cairo, he presented a research paper dealing with the inimitable aspects of geological knowledge contained in the Quran. He said that he did not know what was the state of the art in the field of science during the days of the Prophet Muhammed. But from what we know about the scanty knowledge and means at that time, we can undoubtedly conclude that the Quran is a light of divine knowledge revealed to Muhammed (Peace Be Upon Him).

Here are the concluding remarks of Professor Palmer:

"We need research into the history of early Middle Eastern oral traditions to know whether in fact such historical events have been reported. If there is no such record, it strengthens the belief that Allah transmitted through Muhammed bits of his knowledge that we have only discovered for ourselves in recent times. We look forward to a continuing dialogue on the topic of science in the Quran in the context of geology"[5].

References

1. Nasri Tooba (A Quranic Encyclopedia) by Allamah Sha'raani Islamiah Publishers, Teheran, 1978.
2. Dead Sea: Lowest, Saltiest Point on Earth is home to Masada, by Terry Housholder and Grace Witwer Housholder, http://www.frugalfun.com/.
3. Dead Sea, The lowest point on earth, http://www.meltingpot.fortunecity.com/.
4. A Dictionary of Modern Written Arabic, (Arabic-English) by hans Wehr, McDonald and Evans Ltd. London /see also: Encyclopedia of Quran (Qaamousse Quran, in Persian Language), by ayatollah Ali Akbar Qurashi, Dar Alkutub Al-Islamiah, Teheran, 1999.
5. The Lowest Part on the Face of the Earth, http://www.beconvinced.com/.

Oceanography in Holy Quran

- Oceans and the Internal Waves

Oceans and the Internal Waves

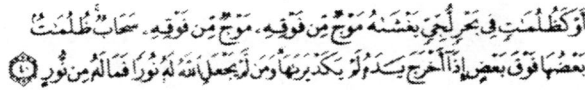

One of the newest discoveries in the field of oceanography is the existence of internal waves in the deep seas and oceans. However, the Quran speaks of this phenomenon in the deep seas:

-Or like utter darkness in the deep sea. There covers it a wave above which is another wave, above which is a cloud, (layers of) utter darkness one above another; when he holds out his hand, he is almost unable to see it; and to whomsoever Allah does not give light, he has no light[1].

Chapter24: verse40

This verse points out to the oceans' internal and superficial waves. The most roaring and horrible waves are those which move surreptitiously underneath the oceans. In the past, the ships, which voyaged to the North Pole, encountered grave difficulties. Now, it is discovered that those problems have been because of the internal waves. In 1900, the Scandinavian mariners set forth the theory of oceans' underneath waves.

Some years ago, a story came out of Toronto about a man who was in the merchant marine and made his living on the sea. A Muslim gave him a translation of the Quran to read. The merchant marine knew nothing about the history of Islam but was interested in reading the Quran. When he finished reading it, he brought it back to the Muslim and asked, "This Muhammed was he a sailor?" He was impressed at how accurately the Quran describes a storm on a sea. When he was told, "No, as a matter of fact, Muhammed lived in the desert" that was enough for him. He embraced Islam on the spot. He was so impressed with the Quran's description because he had been in a storm on the sea, and he knew that whoever had written that description had also been in a storm on the sea.

Internal waves at interface between two layers of water of different densities. One is dense (the lower one), the other one is less dense (the upper one).

The description of "a wave, over it a wave, over it clouds" was not what someone imagining a storm on a sea to be like would have written; rather, it was written by someone who knew what a storm on the sea was like. This is one example of how the Quran is not tied to certain place and time. Certainly, the scientific ideas expressed in it also do not seem to originate from the desert fourteen centuries ago[2].

Scientific Viewpoint

Internal Waves

Although hidden from sight, the interior of the ocean is just as turbulent as its surface. Roughly 40 meters below the surface, there is an abrupt change in both water density and temperature, respectively called the pycnocline and the thermocline. The pycnocline, therefore, is a gradual interface between two fluids of different density. Disturbances travel along fluid interfaces, and disturbances, which travel along the pycnocline, are called internal waves. Since internal waves are not directly visible, they can only be detected by their surface signature and by direct measurements of the pycnocline or thermocline.

Where the density interface is shallow enough to permit the internal wave crests to interact with the sea surface, the waves can be detected by the resulting increased roughness of the surface ocean. Internal waves are believed to be responsible for a great deal of damage. Large amplitude internal waves can create enormous bending moments in offshore structures, and have been reported to displace oil platforms as much a 200 meters in the horizontal direction and 10 meters in the vertical direction. Also, there has been speculation that the loss of the submarine USS

Thresher in 1969 came from an internal soliton carrying the submarine rapidly deeper than its crush depth. There is no direct evidence for that theory, however.

The evolution of internal waves and interaction of internal waves with other waves and structures in the horizontal plane is not understood and has been studied to only a small degree.

Quranic Viewpoint

Now, let us pay attention to the verse:

-...There covereth him a wave, above which is a wave, above which is a cloud. Layer upon layer of darkness.

This part of the verse explicitly indicates the superficial and internal waves and the evidence of this description is the "deep sea" i.e. the oceans not the ordinary seas.
The other noticeable subject is the utter darkness mentioned in the verse. We know that the sunlight is weakened underneath the oceans, because of the refraction and absorption of light, so that in the depth of 1000 m, there scarcely is a light. In the case of cloudy sky, surely its utter darkness is enhanced. Thus, the purport of the verse is clarified while saying:

-When he (the person inside the ocean) holdeth out his hand he scarcely can see it!

Professor Dorja Rao, a specialist in Marine Geology and currently teaching at King Abdul-Aziz University in Jeddah spoke with Sheik Abdul Majid Zendani, a Professor of Islamic Studies in King Abdulaziz University in Jeddah, Saudi Arabia. The sheik relates about his meeting with the professor Rao and that he presented a number of Quranic verses containing scientific signs in the Quran. He was astonished with what he saw and heard. He has read the interpretations of the Quran and its verses in some specialized books. Among these verses, he discussed what Allah said in the Quran:

-Or (the unbelievers state) is like the depths of darkness in a vast deep ocean, overwhelmed with billow topped by billow, topped by (dark) clouds: Depths of darkness, one above another: If a man stretches out his hand, he can hardly see it! For any to whom Allah does not give light, there is no light.

Professor Rao confirmed that scientists now know darkness by means of submarines that have enabled them to dive into the depths of the ocean, where human beings are not able to dive unaided for more than twenty to thirty meters.

Those who dive for pearls do so in shallow waters and can not dive any deeper than this. Human beings can not survive in the deep dark part of the oceans, such as at a depth of 200 meters. But this verse speaks about a phenomenon found in very deep oceans. Hence, the statement of Allah: The darkness in a vast deep sea does not refer to just any sea, because not every sea can be described as having accumulated darkness layered one over another. This sort of layered darkness in deep seas has two causes out of which are the results of the successive disappearances of color, one layer after the other. The light ray is composed of seven colors, and when the light ray hits water, it is scattered into these seven colors.

Between 3 and 30 percent of the sunlight is reflected at the sea surface. Then almost all of the seven colors of the light spectrum are absorbed one after another in the first 200 meters, except the blue light.

If a diver would dive to a depth of 30 meters and gets wounded there, he would not be able to see his blood, because the red color does not reach this depth. In the same way, orange rays are absorbed next. Then at the depth of 50 meters yellow rays are absorbed. At the depth of 100 meters green rays are absorbed. At depths beyond 200 meters, blue rays are absorbed, and so on. From this, we can see that the ocean becomes progressively darker, that is darkness takes place in layers of light. As for the second reason, darkness originates as a result of barriers, which conceal light.

The light rays originate from the sun and absorbed by clouds, which in turn scatter some of the light rays, thus resulting in a layer of darkness under the clouds. This is the first layer of darkness. Then when light rays reach the surface of the ocean, they are reflected by the wave surface, thus giving it a shiny appearance. It is for this reason that when there are waves, the intensity of this reflection depends on the angle of the waves. Therefore, it is the waves that reflect light and therefore cause darkness. The unreflected light rays penetrate into the depth of the ocean, and thus we divide the ocean into two main layers, the surface and the deep part. The surface of the ocean is characterized by the light and warmth, whereas the deep is characterized by darkness.

These two parts of the ocean differ with respect to their properties and characteristics, and the surface is further separated from the deep part by waves. These internal waves were only discovered in the year 1900. Scientists have recently discovered that there are internal waves which "occur on dense interfaces between layers of different densities. The internal waves cover the deep waters of seas and oceans because the deep waters have a higher density than the waters above them. Internal waves act like surface waves. They can also break just like surface waves. Internal waves cannot be seen by the human eye, but they can be detected by studying temperature or salinity changes at a given location.

Underneath these waves which separate the two parts of the ocean, the darkness begins. Fish at these depths cannot see. Their only source of light is from their bodies. This darkness which is layered or tiered one over the other is what is referred to in the Quran:

-Darkness in a vast deep ocean, overwhelmed by billow topped by billow.

Chapter24: verse40

In other words, above those waves there are still more tiers of waves. The latter being found on the surface of the ocean. The Quran then informs us that darkness is:

-Topped by clouds: Depths of darkness, one above another.

Chapter24: verse40

This darkness is caused by the barriers explained, in addition to the darkness caused by the absorption of colors at the different levels that are layered one over the other. The Quran goes on to say:

-When a man stretches out his hand, he can hardly see it! For any to whom Allah gives not light, there is no light.

Chapter24: verse40

Here is total darkness, submarines must bring their source of light with them, so who could have informed Prophet Muhammed of this?

Professor Rao when was presented with many verses dealing with his area of specialization was asked: "What do you think of the existence of scientific information in the Quran? How could Prophet Muhammed have known about these facts 14 centuries ago?"

Professor Rao replied: "It is difficult to imagine that this type of knowledge existed at the time around 1400 years back. Maybe some of the things they have simple ideas about such, but to describe those things in great detail is very difficult. So, this is definitely not a simple human knowledge. A normal human being cannot explain this phenomenon in that much detail. So I thought the information must have come from a supernatural source."

Yes, the source of such knowledge must be from a level beyond that of man. It cannot be from nature, as Professor Rao said, but this is far beyond nature, and far beyond human capability. What Professor Rao was trying to say is that this is something which cannot be attributed to a natural being, for it is truly the speech of the one who knows nature, the universe and its secrets, as the Quran tells us:

-Say: The (Quran) was sent down by Him who knows the secrets (that is) in the heavens and the earth

Chapter25: verse6

It is from Allah. In this way, the testimonies of the scientists are concentrated one after the other in order to prove that this guidance and light contains in it the indisputable evidence of its truth, for the Quran is the source of guidance until the Last Hour.

The miracle of the Quran is an everlasting one, which is renewed until the Last Hour, and which may be known to all people in spite of their different cultural levels and historical times. The Bedouin in the desert as well as the university Professor will find in the Quran that which will suffice him[3].

The land in which Holy Quran was revealed had no resources or inclinations to explore the depths of the oceans. If we assume that Muhammed had seen the seas in his adolescence (this never happened), the domain of his perceptions would have never been beyond the red sea or the Mediterranean seaside and this also does not adapt to what Quran has described about the deep seas. Certainly, these are all vivid proofs for divinity of Quran as a revelation from God Almighty for those who ponder.

References

1. Holy Quran translated by M.H. Shakir, Ansarian Publication, Qum, the Islamic Republic of Iran, 1993.
2. The Amazing Quran, by Dr. Gary Miller, http://www.users.erols.com/.
3. Quran on the seas and Internal Waves, http://www.islam-guide.com/.

Zoology in Holy Quran

- Intelligent World of Animals

Intelligent world of Animals

وَمَا مِن دَآبَّةٍ فِى ٱلْأَرْضِ وَلَا طَٰٓئِرٍ يَطِيرُ بِجَنَاحَيْهِ إِلَّآ أُمَمٌ أَمْثَالُكُم مَّا فَرَّطْنَا فِى ٱلْكِتَٰبِ مِن شَىْءٍ ثُمَّ إِلَىٰ رَبِّهِمْ يُحْشَرُونَ ۝

When we think of animals as being intelligent, we usually conjure up images of animals that have been trained by humans — seeing-eye dogs helping the blind, dolphins performing in shows or apes using sign language. We wonder whether animals think as we do. Nevertheless, intelligence can take different forms, and animals exhibit their problem-solving abilities in many ways every day. They can use tools — chimpanzees use sticks to capture tasty termite snacks. They can communicate with one another — think of the singing humpback whale or the waggle-dancing honeybee. They can interact socially — baboons groom one another and strategically team up to battle a rival.

Although most people consider themselves far superior to the rest of the animal kingdom when it comes to intellectual capacity, in actuality there are many things that, through a combination of instinct and learning, animals can do much better than humans. Could you build a bird's nest or a spider's web? How well could you anticipate the escape behaviors of a rabbit or deer in the heat of a chase? How would you handle navigating thousands of miles between a summer and winter home without maps or road signs, like a migratory bird? We may not fully appreciate animal intelligence until we consider the specific challenges a particular kind of animal faces in day-to-day life, such as finding food, a mate or a home.

The Quran describes the animals and the birds as "peoples" or "nations" just like humans through the verse38 of chapter6:

-There is not an animal in the earth nor a flying creature flying on two wings, but they are peoples like unto you. We have neglected nothing in the book. Then unto their Lord they will be gathered[1].

Chapter6: verse38

This wonderful description of the animals implies that they are similar to us in many ways; namely, they have wisdom by which they contrive their affairs etc. there is a newfound awareness and appreciation in science for their particular unique talents. The scientists clarify that the animal groups have a social and firm relation with each other. The living environment, most notably as in ants and honeybees is in the form of a highly organized community. In addition, there are languages among some of the animal groups by which their members inculcate their intentions to each other.

Former scientists did not believe in existence of wisdom and intelligence in animals. They supposed that they are the living creatures, which have only the senses without the power of thinking.

The behaviors in animals were thought to be only instinctive and inspirational not to be originated from a higher order thinking power. This theory was dominant until the late centuries. Even the famous philosopher "De Carte" stated that animals do not have wisdom and sagacity to perceive, in the way human can think! This assumption about animals was taken into consideration by other scientists, until Darwin, the famous British scientist announced that animals have the power of thinking but in a lower level relative to human being. However, this is the fact that Quran miraculously revealed it fourteen hundred years ago through this glorious verse:

-There is not an animal in the earth nor a flying creature flying on two wings, but they are peoples like unto you...

Chapter6: verse38

Scientific Viewpoint

Animal Behavior

I. Introduction

Animal Behavior here is defined as the vast array of manners in which animals behave. This has fascinated inquiring minds since at least the time of Plato and Aristotle. Particularly intriguing has been the ability of simple creatures to perform complicated tasks-weave a web, build a nest, sing a song, find a home, or capture food-at just the right time with little or no instruction. Such behavior can be viewed from two quite different perspectives, discussed below: Either animals learn everything they do (from "nurture"), or they know what to do instinctively (from "nature"). Neither extreme has proven to be correct.

II. Nurture: The Behaviorists
Until recently the dominant United States school in psychological theory has been behaviorism, whose best-known figures are J. B. Watson and B. F. Skinner. Strict behaviorists hold that all behavior, even breathing and the circulation of blood, according to Watson, is learned; they believe that animals are, in effect, born as blank slates upon which chance and experience are to write their messages. Through conditioning, they believe an animal's behavior is imprinted. Behaviorists recognize two sorts of conditioning: classical and operant.

In the late 19th century the Russian physiologist Ivan Pavlov discovered classical conditioning while studying digestion. He found that dogs automatically salivate at the sight of food-an unconditioned response to an unconditioned stimulus, to use his terminology. If Pavlov always rang a bell when he offered food, the dogs began slowly to associate this irrelevant (conditioned) stimulus with the food. Eventually the sound of the bell alone could elicit salivation. Hence, the dogs had learned to associate a certain cue with food. Behaviorists see salivation as a simple reflex behavior, something like the automatic reflex doctors trigger when they tap a patient's knee with a hammer.

The other category, operant conditioning, works on the principle of punishment or reward. In operant conditioning a rat, for example, is taught to press a bar for food by first being rewarded for facing the correct end of the cage, next being rewarded only when it stands next to the bar, then only when it touches the bar with its body, and so on, until the behavior is shaped to suit the task. Behaviorists believe that this sort of trial-and-error learning, combined with the associative learning of Pavlov, can serve to link any number of reflexes and simple responses into complex chains that depend on whatever cues nature provides. To an extreme behaviorist, then, animals must learn all the behavioral patterns that they need to know.

III. Nature: The Ethologists
In contrast, ethology-a discipline that developed in Europe but that now dominates United States studies as well holds that much of what animals know is innate (instinctive). A particular species of digger wasp, for ex-

ample, finds and captures only honeybees. With no previous experience a female wasp will excavate an elaborate burrow, find a bee, paralyze it with a careful and precise sting to the neck, navigate back to her inconspicuous home, and, when the larder has been stocked with the correct number of bees, lay an egg on one of them and seal the chamber. The female wasp's entire behavior is designed so that she can function in a single specialized way. Ethologists believe that this entire behavioral sequence has been programmed into the wasp by its genes at birth and that, in varying degrees, such patterns of innate guidance may be seen throughout the animal world.

Extreme ethologists have even held that all novel behaviors result from maturation-flying in birds for example, which requires no learning but is delayed until the chick is strong enough-or imprinting, a kind of automatic memorization discussed below.

Programmed Learning

The fourth contribution ethology has made to the study of animal behavior is the concept of programmed learning. Ethologists have shown that many animals are wired to learn particular things in specific ways at preordained times in their lives.

IV. Complex Behavior Patterns

Evolution, working on the four general mechanisms described by ethology, has generated a nearly endless list of behavioral wonders by which animals seem almost perfectly adapted to their world. Prime examples are the honeybee's systems of navigation, communication, and social organization.

Bees rely primarily on the sun as a reference point for navigation, keeping track of their flight direction with respect to the sun and factoring out the effects of the winds that may be blowing them off course. The sun is a difficult landmark for navigation because of its apparent motion from east to west, but bees are born knowing how to compensate for that. When a cloud obscures the sun, bees use the patterns of ultraviolet polarized light in the sky to determine the sun's location. When overcast conditions obscure both sun and sky, bees automatically switch to a third navigational system based on their mental map of the landmarks in their home range.

Study of the honeybee's navigational system has revealed much about the mechanisms used by higher animals. Homing pigeons, for instance, are now known to use the sun as their compass; they compensate for its apparent movement, see both ultraviolet and polarized light, and employ a backup compass for cloudy days. The secondary compass for pigeons is magnetic. Pigeons surpass bees in having a map sense as well as a compass as part of their navigational system. A pigeon taken hundreds of kilometers from its loft in total darkness will nevertheless depart almost directly for home when it is released. The nature of this map sense remains one of ethology's most intriguing mysteries.

Honeybees also exhibit excellent communication abilities. A foraging bee returning from a good source of food will perform a "waggle dance" on the vertical sheets of honeycomb. The dance specifies to other bees the distance and direction of the food. The dance takes the form of a flattened figure 8; during the crucial part of the maneuver (the two parts of the figure 8 that cross) the forager vibrates her body. The angle of this part of the run specifies the direction of the food: If this part of the dance points up, the source is in the direction of the sun, whereas if it is aimed, for example, 70° left of vertical, the food is 70° left of the sun. The number of waggling motions specifies the distance to the food.

The complexity of this dance language has paved the way for studies of higher animals. Some species are now known to have a variety of signals to smooth the operations of social living. Vervet monkeys, for example, have the usual set of gestures and sounds to express emotional states and social needs, but they also have a four-word predator vocabulary: A specific call alerts the troop to airborne predators, one to four-legged predators such as leopards, another to snakes, and one to other primates. Each type of alarm elicits a different behavior. Leopard alarms send the vervets into trees and to the top branches, whereas the airborne predator call causes them to drop like stones into the interior of the tree. The calls and general categories they represent seem innate, but the young learn by observation which species of each predator class is dangerous. An infant vervet may deliver an aerial alarm to a vulture, a stork, or even a falling leaf, but eventually comes to ignore everything airborne except the martial eagle.

The Ant

We are going to attract your attention to a creature that you know quite well, that you meet everywhere without actually giving it much attention, that is highly skillful, highly social and highly intelligent "The Ant". Our aim is to review the lives full of miracles of these minute creatures that are never of any significance in our daily lives.

Technology, collective work, military strategy, advanced communications network, an astute and rational hierarchy, discipline, Perfect City planning... These are fields where human beings may not always be successful enough, but where the ants always are. When you look at these creatures, which are fully armed to defeat tough rivals and to endure the difficult conditions of nature, you may think that all of them are identical.

However, each species of the ant genus - and there are thousands of them - has, in fact, different characteristics. These creatures that have the highest population in the world may open up new horizons for us within the framework of the characteristics referred to above. We shall witness the things these ant communities succeed with their tiny bodies and witness that there is absolutely no difference between their fossils, the oldest of which is about 80 million years old, and their counterparts living today, that run to approximately 8800 species.

The living beings that have the densest population in the world are the ants. For every seven hundred million ants that come into this world, there are only 40 newborn human beings. There is a lot of other amazing information to learn about these creatures.

The ants, one of the most "social" groups among the insect genus, live as societies called "colonies", which are extremely well "organized." Their organization is of such an advanced order that it may be said that in this respect they have a civilization similar to that of humans.

The ants care for their babies, protect their colonies and fight as they produce and store their food. There are even colonies that do "tailoring", that deal in "agriculture" or "animal husbandry". These animals, with their very strong communication network, are so superior as not to be compared to any other organism, with respect to social organization and specialization.

In our day, there are some researchers who are applying their superior intelligence and education working day and night in think tanks formed to formulate successful social organizations and to find lasting solutions to social and economic problems. Ideologues have been producing social models for centuries. Yet when we look at the world in general, no ideal socio-economic social order has so far been reached, in spite of all these intensive efforts. Since the concept of order in human societies has always been based on competition and individual interests, a perfect social order has never been possible. The ants on the other hand, have perpetuated the social system that is ideal for them for millions of years right down to the present day. Then how can these minute creatures form such an order? An answer must certainly be sought for this question.

Evolutionists, when trying to answer this question, claim that ants evolved 80 million years ago from "Tiphiidae", which is an archaic genus of wasps, and that they started socializing 40 million years ago - suddenly, "at their own discretion" - and that they constitute the highest level of the evolution of insects. However, they do not in any way explain the causes and the process of development of this socialization.

The basic mechanism of evolution requires living beings to fight with each other to the end, for their survival. Therefore, each genus and every individual within that genus can think of only itself and its own offspring

(Why and how it started thinking of its offspring is another dead end for Evolution, but we are skipping this point for now). It is, of course, unanswered how this type of a "law of evolution" can form a social system with sacrifice right at its core.

The questions to be answered are not limited to these. Could these creatures whose nerve cells for one million of them only weigh 20 grams, have adopted the resolution to socialize in groups "just like that"? Or could they have got together to set the rules for this socializing after adopting such a resolution? Even if we accept that they could, would all of them obey this new system without exception? Have they formed an advanced social order by founding colonies with millions of members after overcoming all these seeming impossibilities?

How did a "caste system" emerge out of this struggle? First, this question has to be answered: How has the difference between the queen and the worker developed?

Evolutionists at this point will say that a group among the workers abandoned working and developed a physiology different from the worker ants by going through genetic variations over a long period of time. However, we are then faced with the question of how the said "would be queens" were nourished throughout this transformation period. The queen ants do not look for food. They are fed with food brought by the workers. Some workers may have seen themselves as "queens", so how and why have other workers accepted this hierarchy? Furthermore, why have they consented to feed this queen? The "struggle for life" that they are in, according to "evolution", requires that they only think of themselves.

All insects spend most of their time in looking for food. They find and they eat food, then they get hungry again and go off to find more food. They also run from danger. When we accept evolution, we also have to accept that the ants too lived "individually" once upon a time, but that one day, millions of years ago, they decided to become socialized.

The question then arises as to how they "decided" "to form" this social order without any common communication between themselves, because, according to evolution, communication is a consequence of socializing. Furthermore, the question of how they have developed the genetic mutation required for this socialization has no scientific explanation whatsoever.

All these arguments take us to a single point: To claim that the ants started "socializing" one day millions of years ago is to break all the basic rules of logic. The only possible explanation is as follows: The social order was created along with the ants; and this system has not varied since the first ant colony on earth, until today.

Honeybee

The honeybee is so intrinsically a social insect that can survive only as a member of a community, or colony. The colony inhabits an enclosed cavity, its nest. Domesticated colonies are kept in artificial containers, usually wooden boxes, known as hives.

The honey bee community consists of three structurally different forms-*the queen* (reproductive female), *the drone* (male), and the *worker* (non-reproductive female*).* These castes are associated with different functions in the colony; each caste possesses its own special instincts geared to the needs of the colony.

The queen is the only sexually productive female in the colony and thus is the mother of all drones, workers, and future queens. Her capacity for laying eggs is outstanding; her daily output often exceeds 1500 eggs, the weight of which is equivalent to that of her own body.

Anatomically, the queen is strikingly different from the drones and workers. Her body is long, with a much larger abdomen than a worker bee. Her mandibles, or jaws, contain sharp cutting teeth, whereas her offspring have toothless jaws. The queen has a curved, smooth stinger that she can use repeatedly without endangering her own life.

In contrast, the worker honeybees are armed with straight, barbed stingers, so that when a worker stings, the barbed, needle-sharp organ remains firmly anchored in the flesh of its victim. In trying to withdraw the stinger, the bee tears its internal organs and dies shortly thereafter.

The queen bee lacks the working tools possessed by worker bees, such as pollen baskets, beeswax-secreting glands, and a well-developed honey sac. Her larval food consists almost entirely of a secretion called royal jelly that is produced by worker bees. The average life span of the queen is one to three years.

The Worker Bee

Worker bees are the most numerous members of the colony. A healthy colony may contain 80,000 worker bees or more at its peak growth in early summer. Workers build and maintain the nest and care for the brood. They build the nest from wax secreted from glands in their abdomen.

The hexagonal cells, or compartments, constructed by the workers are arranged in a latticework known as the comb. The cells of the comb provide the internal structure of the nest and are used for storage of the developing young bees and all the provisions used by the colony. Comb used for storage of honey is called honeycomb.

Workers leave the hive to gather nectar, pollen, water, and propolis, a gummy substance used to seal and caulk the exterior of the nest. They convert the nectar to honey, clean the comb, and feed the larvae, drones, and the queen. They also ventilate the nest and when necessary, defend the colony with their stings. Workers do not mate and therefore can not produce fertile eggs. They occasionally lay infertile eggs, which give rise to drones.

As with all bees, pollen is the principal source of protein, fat, minerals, and vitamins, the food elements essential for the growth and development of larvae of all three castes. Adult bees can subsist on honey or sugar, a pure carbohydrate diet.

Besides gathering and storing food for all the members of the colony, the workers are responsible for maintaining the brood at 33.9° C (93° F), the optimum temperature required for hatching the eggs and rearing the young. When the nest or hive becomes too hot the workers collectively ventilate it by fanning their wings. During cool weather, they cluster tightly about the nursery and generate heat. The eggs, which are laid one per cell, hatch in three days. The larvae are fed royal jelly for at least two days and then pollen and nectar or honey. Each of the hundreds of larvae in a nest or hive must be fed many times a day.

For the first three weeks of their adult lives, the workers confine their labors to building the honeycomb, cleaning and polishing the cells, feeding the young and the queen, controlling the temperature, evaporating the water from the nectar until it thickens as honey, and many other miscellaneous tasks. At the end of this period, they function as field bees and defenders of the colony.

The workers that develop early in the season live extremely busy lives, which, from egg to death, last about six weeks. Worker bees reared late in the fall usually live until spring, since they have little to do in the winter except eat and keep warm. Unlike other species of bees, honeybees do not hibernate; the colony survives the winter as a group of active adult bees.

The Drone Bee

Drones are male honeybees. They are born without a stinger, defenseless, and unable to feed themselves-they are fed by worker bees. Drones have no pollen baskets or wax glands and cannot secrete royal jelly. Their one function is to mate with new queens. Mating always takes place while in open-air flight, after which a drone dies immediately.

Early investigators of the mating habits of the honeybee concluded that a queen mates only once in her life. Recent scientific studies, however, have established that she usually mates with six or more drones in the course of a few days. The motile sperm of the drones find their way into a small, sac-like organ, called the spermatheca, in the queen's abdomen. The sperm remain viable in this sac throughout the life of the queen.

Drones are prevalent in colonies of bees in the spring and summer months. As fall approaches, they are driven out of the nests or hives by the workers and left to perish.

A. Reproduction and Development

The queen controls the sex of her offspring. When an egg passes from her ovary to her oviduct, the queen determines whether the egg is fertilized with sperm from the spermatheca. A fertilized egg develops into a female honeybee, either worker or queen, and an unfertilized egg becomes a male honeybee, or drone.

The queen lays the eggs that will develop into more queens in specially constructed downward-pointing, peanut-shaped cells, in which the egg adheres to the ceiling. These cells are filled with royal jelly to keep the larvae from falling and to feed them.

Worker bees are raised in the multi-purpose, horizontally arranged cells of the comb. Future workers receive royal jelly only during the first two days, compared to future queens, who are fed royal jelly throughout their larval life. This difference accounts for the great variation in anatomy and function between adult workers and queens. On average, the development of the queen from egg to adult requires 16 days; that of the worker, 21 days; and that of the drone, 24 days.

B. Activities

Field honeybees collect flower nectar. On entering the hive with a full honey sac, which is an enlargement of the esophagus, the field bee regurgitates the contents into the mouth of a young worker, called the house, or nurse, bee. The house bee deposits the nectar in a cell and carries out the tasks necessary to convert the nectar to honey. When the honey is fully ripened, the cell is sealed with an airtight wax capping. Both old and young workers are required to store the winter supplies of honey.

Pollen is carried into the nest or hive on the hind legs of the field bees and placed directly in the cells. The pollen of a given load is derived mostly from plants of one species, which accounts for the honeybee's outstanding role as pollinator. If it flew from one flower species to another, it would not be effective in the transfer of pollen, but by confining its visits on a given trip to the blossoms of a single species, it provides the cross-pollination required in many varieties of plants.

C. Communication

An amazing symbolic communication system exists among honeybees. In studies of bees begun in the early 1900s, the Austrian zoologist Karl von Frisch determined many of the details of their means of communication. In a classic paper published in 1923, von Frisch described how after a field bee discovers a new source of food, such as a field in bloom, she fills her honey sac with nectar, returns to the nest or hive, and performs a vigorous but highly standardized dance. If the new source of food is within about 90 m (about 295 ft) of the nest or hive, the bee performs a circular dance, first moving about 2 cm (about .75 in) or more, and then circling in the opposite direction. Numerous bees in the nest or hive closely follow the dancer, imitating her movements. During this ceremony, the other workers scent the fragrance of the flowers from which the dancer collected the nectar. Having learned that food is not far from the nest or hive, and what it smells like, the other bees leave the nest or hive and fly in widening circles until they find the source.

If the new source of nectar or pollen is farther away, the discoverer performs a more elaborate dance characterized by intermittent movement across the diameter of the circle and constant, vigorous wagging of her abdomen.

Every movement of this dance seems to have significance. The number of times the bee circles during a given interval informs the other bees how far to fly for the food. Movement across the diameter in a straight run indicates the direction of the food source. If the straight run is upward, the source is directly toward the sun. Should the straight run be downward, it signifies that the bees may reach the food by flying with their backs to the sun. In the event the straight run veers off at an angle to the vertical, the bees must follow a course to the right or left of the sun at the same angle that the straight run deviates from the vertical. Bees under observation in a glass hive demonstrate their instructions so clearly that it is possible for trained observers to understand the directions given by the dancers.

Certain aspects of the dance language, such as how attendant bees perceive the motion of dancers in the total darkness of the nest or hive, are still unknown. The dance language is an important survival strategy that has helped the honeybee in its success as a species.

D. *Problems of Survival*

Honeybees are subject to various diseases and parasites. American and European foulbrood are two widespread contagious bacterial diseases that attack bee larvae. A protozoan parasite, Nosema, and a virus cause dysentery and paralysis in adult bees. Two species of blood-sucking parasitic mites are particularly troublesome for beekeepers and are currently affecting wild honeybees worldwide. The honey bee tracheal mite lives in the breathing tubes of adult bees; the varroa mite lives on the outside of larvae and adults. These mites have killed tens of thousands of honeybee colonies in North America during the past ten years. Scientific breeding programs are attempting to develop tolerant strains of domestic honeybees to replace the mite-susceptible ones currently used. Tracheal mite infestations can be reduced by fumigation of the hive with menthol fumes. Varroa mites are controlled with a miticide or, in some European countries, with fumes of formic acid. Certain hive management techniques also can reduce infestations.

Many other animals prey upon individual honeybees, which may sometimes weaken colonies. Examples are cane toads and bee-eaters (birds), which pick off foragers near the colony entrance; robber flies, which take individual foragers as they visit flowers; and hornets and bee wolves (wasps), which may enter the nest or hive and steal larvae. Bears have an insatiable appetite for honey and bee larvae and may destroy many nests or hives in a single raid.

Honey bee colonies used in commercial pollination and those kept in urban areas are exposed to pesticides, fungicides, fertilizers, and other agricultural chemicals and are frequently poisoned by accident. This is a major concern of modern beekeepers.

Honeybees have become the primary source of pollination for approximately one-fourth of all crops produced in the United States and some other countries. The value of the crops that rely on such pollination has been estimated as high as $10 billion annually in the United States. Examples of fruit crops that rely on honeybees are almonds, apples, apricots, avocados, blackberries, blueberries, cantaloupes, cherries, cranberries, cucumbers, pears, raspberries, strawberries and watermelons. The seeds of many vegetables are also produced with honey bee pollination; examples include alfalfa, asparagus, broccoli, brussels sprouts, cabbage, carrots, clover, cotton, cucumbers, onions, radishes, squash, sweet clover, and turnips.

Many species of wild pollinators have disappeared from the land as their habitats have been destroyed or altered by humans. The honeybee has taken over as pollinator of many of the wild plants that remain; its ecological value in this regard is tremendous.

Honeybees are the sole source of honey and beeswax, a fine wax with unusual qualities. Honeybees also produce propolis, a gummy substance made from tree sap that has antibacterial properties, and royal jelly and pollen for human consumption. Honeybee venom is extracted for the production of antivenom therapy and is being investigated as a treatment for several serious diseases of the muscles, connective tissue, and immune system, including multiple sclerosis and arthritis.

Quranic Viewpoint

So far, we generally looked through the animal behaviors from different aspects of nature, reproduction and development, communication, activities, survivals etc in detail. Also, we particularly studied the ants and honeybees' wonderful world and how they are organized in very developed groups. We perceived that they amazingly live in an intelligent world, in which they learn, communicate and develop in the ways like humans do. This is what the Quran inspirationally have stated 1400 years ago:

-There is not an animal in the earth nor a flying creature flying on two wings, but they are peoples like unto you. We have neglected nothing in the book. Then unto their Lord they will be gathered.
Chapter6: verse38

It would be interesting to recount the anecdote of prophet Solomon and his communication and conversation with the bird and jinns (demons) which is stated in the Quran, chapter 27 "The ant":

-And there were gathered together unto Solomon his armies of the jinn and humankind, and of the birds, and they were set in battle order.
Chapter27: verse17

-Till, when they reached the valley of the Ants, an ant exclaimed: O ants! Enter your dwellings lest Solomon and his armies crush you, unperceiving.
Chapter27: verse18

-And (Solomon) smiled, laughing at her speech, and said: My Lord, arouse me to be thankful for Thy favor wherewith Thou hast favored me and my parents, and to do good that shall be pleasing unto Thee, and include me in (the number of) Thy righteous slaves.
Chapter27: verse19

-And he sought among the birds and said:
How is it that I see not the hoopoe, or is he among the absent?
Chapter27: verse20

-I verily will punish him with hard punishment or I verily will slay him, or he verily shall bring me a plain excuse.
Chapter27: verse21

-But he was not long in coming, and he said: I have found out (a thing) that thou apprehendeth not, and I come unto thee from sheba

with sure tidings

Chapter27: verse17

The worker ants are female. The Quran uses the feminine gender when talking about this ant. The word "Namlah" is the female form of the word "Naml" meaning "the female ant". Another knowledge comes from the Quran, which it was not discovered before Muhammed's time (peace be upon him).

When we look at ants, we see that things are no different for them either. Allah has inspired in them a social order also and they abide by it absolutely. This is the reason why each group of ant performs the duty assigned to it perfectly with absolute self-surrender and does not strive for more.

The Female Bee in the Quran

In the 16th chapter, the Quran mentions that the female bee leaves its home to gather food. One might guess the bee they see is female." Certainly, he has a one in two chance of being right. So it happens that the Quran is right. However, it also happens that that was not what most people believed at the time when the Quran was revealed.

Even upon close inspection, one cannot tell the difference between a male and a female bee. It takes a specialist to do that, but it has been discovered that the male bee never leaves his home to gather food.

Historically people thought bees to be male. In Shakespeare's play, Henry the Fourth, some of the characters discuss bees and mention that the bees are soldiers and have a king. That is what people thought in Shakespeare's time - that the bees that one sees flying around are male bees and that they go home and answer to a king. However, that is not true at all. The fact is that they are females, and they answer to a queen. Yet it took modern scientific investigations in the last 300 years to discover that this is the case.

When mentioning the bees who have a social order similar to that of the ants, Allah states in the Quran that this social order has been "revealed" to them:

-And your Lord revealed to the bee: "Build dwellings in the mountains and the trees and also in the structures which men erect. Then eat from every kind of fruit and travel the paths of your Lord, which have been made easy for you to follow." From inside them comes a drink of varying colors, containing healing for mankind. There is certainly a sign in that for people who reflect.

Chapter16: verse68-69

As we explore the special world of animals, this perfect system will earn our admiration and increase the need for thinking and investigating. At the same time, we shall witness Allah's immaculate creation, which is a tremendously important work. In the Quran, the type of person who thinks about nature and thus recognizes the omnipotence of Allah is praised as a model for those who believe.

As we have depicted, no contradiction exists between the preceding verses of the Quran and the respective scientific facts. Can any other religion offer its adherents such comfort in knowledge and faith? The presence of such an amazing information in the Quran could not be of an ordinary matter as in that time, the human knowledge concerning the animal behavior and intelligence was not sufficient enough especially for the unlettered Muhammed (peace be upon him) to compose these wonderful statements. This may cause us to think more deeply on and to feel admiration for the superior power and unequaled art of creation of Allah, Who has made all things.

References
1. Holy Quran translated by Marmaduke Pickthall, George Allen and Unwin Ltd., London, Fifth Edition 1969.

Scientific Sources
- Animal Behavior, Encarta Encyclopedia 2001
  http://encarta.msn.com, 2001 Contributed by Carol Grant Gould, Ph.D.
- Research Associate, Department of Biology, Princeton University. Contributor to Harper's, The Sciences, New Scientist, Science '81, and other publications. James L. Gould, B.S., Ph.D., Professor, Department of Ecology and Evolutionary Biology, Princeton University. Co-author of In Search of the Animal Mind, Chance and Causation, and other books.
- The Amazing Quran/Female Bee, by Dr. Gary Miller, http//:www.users.erols.com.
- The Miracle in the Ant, by Harun Yahya, http//:www.harunyahya.com.

Bibliography

1. Holy Quran translated by Marmaduke Pickthall, George Allen and Unwin Ltd., London, Fifth Edition 1969.
2. Holy Quran translated by Arthur J. Arberry, Ansarian Publication, Qum, the Islamic Republic of Iran, 1993.
3. Holy Quran translated by Abdullah Yusuf Ali, Dar Al Arabia Publication, Beirut, Lebanon, 1968.
4. Holy Quran translated by M.H. Shakir, Ansarian Publication, Qum, the Islamic Republic of Iran, 1993.
5. Holy Quran translated by Thomas Ballantine Irving, Suhrawardi Research and Publication Center, Teheran, 1998.
6. Holy Quran translated by Muhammed Mahdi Fulaadvand (in Persian), Dar Al-Quran Al-Kareem, Teheran, 1995.
7. Quran and the Last Prophet (in Persian), by Ayatollah Makarim Shirazi, Dar AL-Kotob Al-Islamiah Publishers, Qum, the Islamic Republic of Iran, 1996.
8. Quran and the New Medicine (in Persian), by Doctor M. Noori Shadkaam, Dar Al-kotob Al-Islamiah Publishers, Qum, the Islamic Republic of Iran, 1996.
9. Lissaan Al-Arab, an Arabic-Arabic Dictionary, by Ibn Manzoor, Dar Sader Publishers, Beirut, 1997.
10. Travel to Space, a series of "Quran and Nature" (in Persian), by Abdulkareem Bi'azaar Shirazi, Be'sat Publishers institute, Teheran, 1971.
11. The Bible, The Quran and Science (Le Bible, le Coran et la Science)," The Holy Scriptures examined in the light of modern knowledge, by Dr. Maurice Bucaille, French physician, Seghers, Paris, 1987, English

version published by North American Trust Publication, 1978.
12. Majma' Al-bayan, An Interpretation of Quran (in Persian), by Abu-Ali Tabarasi, Farahani Publishers, Teheran, 1979.
13. Nasre Tooba (A Quranic Encyclopedia) by Allamah Sha'raani, Islamiah Publishers, Teheran, 1978.
14. Ahsan Al-Hadith, an Encyclopedia of Holy Quran, by Ali Akbar Qurashi.
15. Al-Meezaan, an Interpretation of Quran, (in Persian) Vol. 40, by Muhammed Hussein Tabataba'i, Muhammadi Publishers, Teheran, 1982.
16. Quran's Miracles in the era of Computer and Space, by Professor Dr. Rashid Khalife.
17. Miracle of the Quran (in Persian), by Allaami Tabataba'i, Raja Cultural Publishing Center, Teheran, 1984.
18. Knowledge of the Quran (in Persian), by Mustafa Asraar, Rouyaan Publication, Teheran, 1996.
19. Manifestation of the Quran in Hadith (in Persian), by Morteza Khosravi, Air Force Publishers, Teheran, 1997.
20. Black Holes in Space, by Patrick Moore and Iain Nicolson, Orbach and Chambers Ltd. 1979.
21. Aghrab Al-mawarid, an Arabic-Arabic Dictionary by Allamah Sa'eed Sa'eed Al-Khouli Al-Shartoon Al-Lubnaani.
22. Encyclopedia of Quran (Qaamousse Quran), by Ali Akbar Qurashi, Dar Al-kotob Al-Islamiah, Teheran, 1999.
23. Monjid Al-Tollab (Arabic-Arabic Encyclopedia), by Fu'aad Afraum Al-Bostaani.
24. The Anatomy of the Aeroplane, by Darrol Stinton, BSP Professional Books, UK, 1987.
25. Eyewitness Encyclopedia of Space and the Universe, by Dorling Kindersley Publishers Limited, London, 1990.
26. Exploration of the Universe, by Abell, Morrison, Wolff, Saunders College publishing, USA, 1987.
27. Exploration of the Universe, by George Abell, Hot, Rinehart and Winston Inc., USA, 1969.
28. Al-Mawrid, a Modern English-Arabic Dictionary, by Munir Ba'albaki, Dar El-Ilm Lil-Malayen, Beirut, 1986.
29. A Dictionary of Modern Written Arabic, (Arabic-English) by Hans Wehr, McDonald and Evans Ltd. London.
30. Cambridge International Dictionary of English, Cambridge University Press, 1995.
31. Concepts of Contemporary Astronomy, by Paul Hauge, McGraw-Hill publishers, 1974.
32. Space, Stars, Planets and Spacecraft, by Dorling Kindersley Publishers Limited, London, 1990.
33. The Birth and Death of Stars, by Isaac Asimov, Milwaukee, Gareth

Stevens Publishing, 1989.
34. Philips' Moon Map, Philips Publication, London.
35. Astronomy: From the Earth to the Universe, by Jay M. Pasachoff, Saunders College Publishing, USA, 1991.
36. What's Inside Spacecraft? By Alexandra Parsons, Dorling Kindersley Publishers Limited, London, 1992.
37. Aerodynamics, Aeronautics and Flight Mechanics by McCormick, John Wiley and Sons Inc. 1995.
38. Dynamics of Flight, Stability and Control, by Bernard Etkin, John Wiley and Sons Inc. 1996.
39. The Design of the Aeroplane, by Darrol Stinton, BSP Professional books, UK, 1989.
40. Fundamentals of Aerodynamics, by John D. Anderson Jr., Professor of Aerospace Engineering University of Maryland, McGraw-Hill, Inc., 1991.
41. Shorter Oxford English Dictionary, Oxford University Press, 1993.
42. Webster Comprehensive Dictionary of the English Language, Trident Press international, USA, 1996.
43. The Aryanpur Progressive English-Persian Dictionary, By Dr. Manoochehr Aryanpur-Kashani, Jahan Rayane computer World Co. Teheran, 2001.
44. The Concise Persian-English Dictionary, By Dr. Abbas Aryanpur-Kashani, and Dr. Manoochehr Aryanpur-Kashani, Amir Kabir Publication, Teheran, 1997.
45. The Larger Kamangir English to Persian Dictionary, by Arthur N. Wollaston, Edited by S. Solh-joo, Teheran, 1996.

وَتَمَّتْ كَلِمَتُ رَبِّكَ صِدْقًا وَعَدْلًا لَا مُبَدِّلَ لِكَلِمَاتِهِ وَهُوَ السَّمِيعُ الْعَلِيمُ ﴿١١٥﴾

Perfected is the word of thy Lord in truth and justice. There is naught that can change His words. He is the Hearer, the Knower.

Chapter 6: verse 115

Printed in the United States
By Bookmasters